Clifford Richardson

Methods of analysis of commercial fertilizers, cattle foods, dairy products, sugar and fermented liquors

Clifford Richardson

Methods of analysis of commercial fertilizers, cattle foods, dairy products, sugar and fermented liquors

ISBN/EAN: 9783337201104

Printed in Europe, USA, Canada, Australia, Japan

Cover: Foto ©Andreas Hilbeck / pixelio.de

More available books at **www.hansebooks.com**

U. S. DEPARTMENT OF AGRICULTURE.

DIVISION OF CHEMISTRY.

BULLETIN   No. 19.

METHODS OF ANALYSIS

OF

COMMERCIAL FERTILIZERS, CATTLE FOODS,

DAIRY PRODUCTS, SUGAR, AND FERMENTED LIQUORS,

ADOPTED AT THE

FIFTH ANNUAL CONVENTION OF THE ASSOCIATION OF OFFICIAL
AGRICULTURAL CHEMISTS, HELD AT THE U. S. DEPART-
MENT OF AGRICULTURE AUGUST 9 AND 10, 1888.

EDITED BY

CLIFFORD RICHARDSON,
SECRETARY OF THE ASSOCIATION.

WASHINGTON:
GOVERNMENT PRINTING OFFICE.
1888.

7717—Bull 19

LETTERS OF TRANSMITTAL.

DEPARTMENT OF AGRICULTURE,
DIVISION OF CHEMISTRY,
Washington, D. C., September 4, 1888.

DEAR SIR: Complying with your invitation the Association of Official Agricultural Chemists met in the rooms of the Chemical Division August 9 and 10 this year. I have the honor to submit for your approval, with a view to publication as Bulletin 19 of this division, the official proceedings of that meeting, as attested by the inclosed letter from the secretary.

Respectfully,

H. W. WILEY,
Chemist.

Hon. N. J. COLMAN,
Commissioner of Agriculture.

WASHINGTON, *August* 31, 1888.

DEAR SIR: I have the honor to hand you herewith, for publication, an abstract of the proceedings of the fifth annual convention of the Association of Official Agricultural Chemists, together with the methods adopted for official use during the ensuing year for the analysis of commercial fertilizers, cattle foods, dairy products, sugar and sugar products, and fermented liquors.

To this has been appended a list of the reporters upon the different subjects for 1888–'89, of the officers of the association, and the constitution as amended.

The demand for information of the nature given in the proceedings is rapidly increasing, and the methods of the association are almost universally adopted in this country as standards.

Very respectfully,

CLIFFORD RICHARDSON,
Secretary, Association of Official Agricultural Chemists.

Dr. H. W. WILEY,
Chemist, etc.

3

UNIFORM METHODS FOR THE ANALYSIS OF FERTILIZERS, CATTLE FOODS, DAIRY PRODUCTS, SUGAR, AND FERMENTED LIQUORS.

PROCEEDINGS OF THE FIFTH ANNUAL CONVENTION OF THE ASSOCIA-TION OF OFFICIAL AGRICULTURAL CHEMISTS, HELD AT WASHINGTON, AUGUST 9 AND 10, 1888.

MORNING SESSION, THURSDAY.

In accordance with the call of the secretary, the Association met in the library of the Department of Agriculture at 10 o'clock, the president, Mr. P. E. Chazal, in the chair.

The following members and "others interested in the objects of the Association" were present:

The president, Mr. P. E. Chazal, State chemist of South Carolina.

The vice-president, Dr. W. J. Gascoyne, of Baltimore.

The secretary, Mr. Clifford Richardson, of Washington, D. C.

Of the executive committee, Prof. John A. Meyers, of the West Virginia Agricultural Experiment Station.

Dr. H. W. Wiley, Washington, D. C.; Dr. C. A. Crampton, Washington, D. C.; Prof. William C. Stubbs, of Louisiana; Prof. E. A. von Schweinitz, of Salem, N. C.; Prof. Richard H. Gaines, of Richmond, Va.; Prof. William Frear, of State College, Pa.; Prof. H. J. Patterson, of Agricultural College, Md.; Prof. W. L. Hutchinson, of Agricultural College, Miss.; Prof. G. S. Fellows, of Washington, D. C.; Prof. E. B. Voorhees, of New Brunswick, N. J.; Mr. B. B. Ross, of Louisiana; Mr. Edgar Richards, of Washington, D. C.; Prof. M. A. Scovell, of Lexington, Ky.; Prof. E. H. Farrington, of Hanover, N. H.; Prof. C. W. Dabney, jr., of Knoxville, Tenn.; Prof. F. A. Holton, of Iowa City, Iowa; Mr. W. M. Saunders, of Providence, R. I.; Prof. H. C. White, of Athens, Ga.; Messrs. Knorr, Fake, Edson, and Dugan, of the U. S Department of Agriculture.

Letters of regret were read from Dr. E. H. Jenkins, of Connecticut; Dr. G. C. Caldwell, of New York; Prof. George F. Cook, of New Jersey, and others.

The president, in calling the meeting to order, said that he would omit the usual address, as there was a desire to proceed at once to business, with the view of accomplishing the objects of the convention in as short a time as possible.

On a call for reports of committees, Mr. Richardson, in the absence of the chairman, presented the report of the Committee on Cattle Foods, as follows:

REPORT OF THE COMMITTEE ON CATTLE FOODS.

Your committee offer the following report on the work done during the year on the analysis of fodders and feeding stuffs:

Only six out of a dozen or more who undertook to make these analyses for the committee were able to do the work. In many cases so much time was consumed, in the latter part of the year, in the reorganization of the experiment stations, under the provisions of the Hatch bill, that too little was left for these deferred analyses. Moreover, some of the reports were received so late that no copy of the results sent by all the analysts could be sent to each of the parties doing the work, so that if any one so desired he might repeat the analyses with a view to some possible explanation of the discrepancies that occur.

There is no reason to doubt that these analyses, the results of which are given below, were made by one method and with careful attention to the directions given by your committee, and by analysts at least quite up to the average, in respect to skill and special training, of those who commonly do this sort of work at the experiment stations, and yet there are some very serious differences in the results.

It appears to your committee, therefore, that there is occasion for more work on this line, and it may reasonably be hoped that with the large number of experiment stations in operation now a much larger number of participants in this work can be found who can undertake the analyses and complete them in season for a more careful preparation of the report by the committee having it in charge.

Your committee would, therefore, suggest that, if the work is continued, the samples, three in number, as hitherto, be sent out before the middle of October; that the analyses be made strictly according to one scheme, without, however, excluding alternate methods that any one may wish to try in addition to those prescribed by the committee; that the time for receiving the reports be limited to February 1; that a tabulated copy of all results received be sent out within two weeks thereafter to each contributor, and that then a limited time be allowed, say one month, for any revision of his work that he may wish to make.

As one illustration of such possible revision, the uniformly higher results reported in the first line of each list below, on nitrogen and protein, may be due to some fault in the standardization of the solutions used for the titration of the distilled ammonia solutions. Additional illustrations may occur to others on inspection of the figures.

Only one report was received on any alternate methods. At the Connecticut station all the samples were analyzed also by the method in common use there; but no description of the methods followed accompanied the results. These results given for the air-dried substance were in very many of the cases identically the same as those obtained by the committee method, and in no case was there any serious difference; but as the moisture was in all cases nearly or a little over 1 per cent. higher, they would, if calculated to dry substance, be somewhat higher than those obtained by the committee method.

The following interesting results were given also by the same station on different methods of determining the moisture and the fat in the dried residue:

	Corn fodder.	Bran.	Cotton-seed meal.
Moisture by the committee method	6.56	7.84	6.43
Fat in the above	1.45	4.05	12.05
Moisture at 100° in stream of hydrogen	7.33	8.89	7.74
Fat in the above	1.78	4.00	12.42
Moisture at 110–115° in stream of hydrogen	7.73	9.41	7.03
Fat in the above	1.68	4.25	12.50

Name of substance.	Reported by—	Moisture.	In the dry substance.				
			Ash.	Ether extract.	Fiber.	Crude protein.	N-free extract.
Corn fodder.........	Caldwell	5.80	10.59	2.25	29.25	10.90	47.00
	Frear	6.67	10.34	2.15	33.79	10.59	43.13
	Jenkins	6.56	6.60	1.55	29.90	9.90	52.05
	Wilkinson	11.53	8.51	3.31	31.40	9.79	46.97
	Weber	9.04	8.77	2.18	27.88	9.88	51.29
	Peter	7.04	9.48	1.43	24.11	9.56	58.38
Bran.................	Caldwell	8.45	7.52	4.54	10.86	19.57	57.51
	Frear	8.35	7.34	4.51	9.97	17.78	60.40
	Jenkins	7.64	7.01	4.25	8.40	18.13	62.21
	Wilkinson	11.10	6.39	5.79	9.64	16.90	61.58
	Weber	9.90	6.16	4.99	9.10	17.96	61.09
	Peter	8.46	6.19	4.01	7.71	16.75	66.68
Cotton-seed meal	Caldwell	6.76	9.23	10.68	6.70	50.93	22.46
	Frear	6.98	9.06	13.93	4.25	47.85	24.92
	Jenkins	6.43	8.10	12.88	2.86	49.01	27.13
	Wilkinson	8.37	8.12	14.71	3.56	46.13	27.48
	Weber	7.60	8.01	13.76	3.27	47.55	24.41
	Peter	6.14	7.96	12.28	2.51	46.06	25.05

In the dry substance.

	Corn fodder.		Bran.		Cotton-seed meal.	
	Total.	Albuminoids.	Total.	Albuminoids.	Total.	Albuminoids.
Frear.......................	1.69	1.02	2.85	2.40	7.65	6.20
Jenkins.....................	1.57	1.34	2.91	2.64	7.84	7.64
Peter	1.53	1.21	2.68	2.53	7.37	7.00

Respectfully submitted.

G. C. CALDWELL,
Chairman.
CLIFFORD RICHARDSON,
WM. H. JORDAN,
Committee.

Mr. Richardson said that, in view of the large amount of discrepancy in the results, he believed it advisable to confine future work to one method without alternates. He added that, with the establishment of so many experiment stations during the present year, all of which would be interested in this subject, a very large increase in the number of analysts was to be expected, and the results at the next convention would admit of drawing more definite conclusions. He had found that in the alternate method for determining fiber, recommended last year, the use of bottles could only be made a success when live steam was available. In a water-bath, owing to sudden and unequal heating, cracking seemed unavoidable. Dr. Frear found difficulty in securing concordant results on cotton-seed meal, and regretted the absence of all the analysts but himself, which prevented an interchange of experiences.

On motion of the latter gentleman, the report and recommendations of the committee were adopted.

Dr. Wiley then presented the report of the committee on dairy products, as follows:

REPORT OF THE COMMITTEE ON DAIRY PRODUCTS.

BY H. W. WILEY.

Mr. PRESIDENT: I found it impracticable to secure a meeting of the committee appointed at the last meeting to consider the subject of dairy products, and have therefore no report considered by all of the members of the committee to present. In the absence of such a document I thought it best to submit a brief résumé of the work which has been done in dairy products during the year since our convention last met. The methods proposed at our last meeting for the analysis of dairy products have given fairly good satisfaction, and I therefore recommend that they be continued unchanged for another year. Many points which are now in dispute will have been settled by that time, and at our next annual meeting any necessary changes in methods of analysis can be made.

With this introduction I beg to submit the following abstracts of the progress made in the analysis of dairy products during the year just passed.

MILK ANALYSIS.

USE OF ASBESTOS CLOTH INSTEAD OF BLOTTING PAPER IN THE ADAMS METHOD.

Johnstone has reported to the Society of Public Analysts favorable results attending his experiments in substituting asbestos paper for blotting-paper in the Adams method of fat estimation.—(Abstract from the Analyst, December, 1887, p. 234.)

[NOTE.—I called attention to the use of asbestos paper for this purpose two years ago, as will be seen on page 81 of Bulletin No. 13, Part 1.]

AN INSTRUMENT FOR CALCULATING MILK RESULTS.

Richmond has described an instrument for calculating milk analyses based on the relations existing between fat, solids not fat, and specific gravity in milk as established by the paper of Hehner and Richmond, published in the Analyst, vol. 13, p. 26. The instrument consists of a slide rule on one side of which a scale, of which one division equals 1 inch, represents total solids. The other scale, one division of which equals 1.164 inches, represents the fat; while the scale representing specific gravity has a division equal to 1.254 inches.

The instrument is based upon the formula $T - .254\ G = 1.164\ F$. To use the instrument the lines indicating total solids and specific gravity found by analysis are placed together. The fat is then read off by an arrow on the other side. Where many analyses are to be made the instrument is especially valuable since it eliminates all chances of error in calculation.—(Abstract from the Analyst, 1888, vol. 13, No. 144, p. 65.)

ESTIMATION OF FAT IN MILK AND CREAM.

Werner Schmid proposes the following method of estimating milk fat: A test tube of about 50 cubic centimeters capacity is graduated in tenths of a cubic centimeter; 10 cubic centimeters of milk or 5 cubic centimeters of cream are placed in the tube and 10 cubic centimeters concentrated hydrochloric acid. The contents of the tube are boiled with constant shaking until the liquid is dark brown. After cooling in water add 30 cubic centimeters ether, shake, allow to stand until the ether solution has separated, measure its volume and remove 10 cubic centimeters with a pipette; place in a porcelain crucible, evaporate, and dry at 100° in air-bath. Calculate to total volume of the ether solution. The results are said to be exact.—(Zeit. Anal. Chem., 27, p. 464.)

Hehner and Richmond have made an exhaustive study of the relations existing between the specific gravity, fat, and solids not fat, in milk. Three sets of determinations of the fat were made, viz, extraction from a paper coil, from plaster of Paris, and the direct extraction of the dried total solids. All previous formulæ relating to the above relations have been based upon imperfect methods of fat extraction. The work of Hehner and Richmond is therefore especially valuable on account of being based upon the Adams method of extraction. Forty-two analyses of various kinds of milk were made. For discussion of the formulæ the original paper is cited. The working formula obtained is T — .254 G = 1.164 F.

A table accompanying the paper g'ves the percentage of fat, calculated for specific gravities ranging from 1024.0 to 1034.0, and for total solids from 10 per cent. to 15 per cent., inclusive.

The author strongly recommends that no milk analyses be accepted as correct which do not correspond closely with the calculated results.—(Abstract from the Analyst, vol. 13, No. 142, p. 26.)

SOURCES OF ERROR IN SOXHLET'S METHOD OF ESTIMATING FAT IN MILK.

Weinwurm has determined the error which may arise by allowing the ether fat solution obtained by Soxhlet's method to stand for varying lengths of time before its specific gravity is determined. As a result of his experiments it appears that the apparent percentage of fat is raised by allowing the solution to stand for a long time. For the first twenty-four hours, however, this increase is only small, the amount, at most, being a few hundredths of a per cent. By standing a long time the apparent percentage of fat gradually increases, and appears to reach a maximum at the end of ten days. The amount of increase in that time may be .25 of 1 per cent.—(Abstract from Repertorium No. 19 of the Chemiker-Zeitung, 1888, p. 151, and M. Zg. 1888, 17, p. 401.)

ESTIMATION OF DRY SUBSTANCE AND FAT IN MILK BY MEANS OF WOOD FIBER.

Gantter proposes the use of wood fiber for the estimation of fat in milk, which he uses as follows:

Two grams of wood fiber are dried at 105° to constant weight in a dish containing a small glass rod. The total weight of the dish, glass rod, and fiber is noted. After weighing the dish, which is covered with a neat-fitting cover to prevent the absorption of water, 5 to 6 grams of the milk are poured upon the wood fiber and the exact amount taken determined by reweighing the bottle. During evaporation the mass of fiber is pressed from time to time against the sides of the dish, so that no particles thereof remain attached thereto. The final drying is made in an air-bath, and the whole time required for the evaporation and drying should be about two hours and a half. From the increase in weight the amount of dry substance is determined. The mass of fiber is now transferred to a Soxhlet extraction apparatus, and, if necessary, the dish rinsed with petroleum ether. The extraction of fat and the weighing thereof are carried on in the usual way. The author has also used this method for the determination of water in butter and other fats and oils.—(Abstract from the Zeitschrift für analytische Chemie, vol. 26, No. 6, p. 678).

THE CHEMICAL ACTION OF SOME MICRO ORGANISMS IN MILK.

Warington has studied the effect of certain micro-organisms on the curdling of milk. This curdling is effected either by the formation of a rennet-like ferment or by the production of lactic acid. The amount of lactic acid required to curdle milk depends on the temperature, the amount growing less as the temperature rises. Five different organisms were examined which had the power of curdling milk, but in very different degrees. They were *Staphylococcus candidus: B. termo; M. gelatinosus, B. fluorescens liquescens*, and *M. ureæ.*

Some of them readily curdled milk at a temperature as low as 10°. They do not, however, at that temperature produce an appreciable acidity. They therefore act plainly as a ferment. A large quantity of gas is evolved during the action of the same on milk.

Five organisms were also found to act as peptonisers; they are, *B. subtilis, B. anthracis, B. floccus, B. toruliformis,* and Finkler's comma. The milk treated with these organisms at 22° becomes clear after a few days. The clear fluid is rich in pepton. A few organisms render milk after a time decidedly alkaline; two of these are, *B. fluorescens non liquescens* and the *Bacillus of septicæmia.*

Cultivation in milk is an excellent method of distinguishing micro-organisms.—(Abstract from the Chemical News, June 22, 1888.)

REDUCTION AND ROTATION POWER OF MILK SUGAR.

Deniges and Bonnans have reviewed the literature concerning the reduction of copper solution by milk sugar. As compared with pure glucose the reducing power of lactose is as 96 to 136. The copper solution employed contained 34.65 grams of pure crystallized copper sulphate and 10 cubic centimeters pure sulphuric acid per liter. The alkaline solution was composed of 250 cubic centimeters of soda lye of 36° strength, in which was dissolved 150 grams of crystallized Rochelle salts. Ten cubic centimeters of this solution was added to each 10 cubic centimeters of the copper solution.

For the determination of the rotatory power of lactose the author used a pure sugar, obtained by three crystallizations. The specific rotatory power for the anhydrous sugar at 20° was 55.30 for concentrations varying from 4 to 36 per cent.; for the crystallized sugar the specific rotatory power in the same condition is 52.53. The general formula for the crystallized salt would be $[a] = 52.53 + (20 - T) \times .055$. The number .055 is the correction to be applied for each variation of 1° in temperature. The numbers obtained by the authors are exactly those given by Schnöger.—(Abstract taken from the Journal de Pharmacie et de Chimie, April and May, 1888, pp. 363 and 411.)

ESTIMATION OF SUGAR IN MILK BY THE POLARISCOPE.

Vieth discusses the method adopted by the Association of Official Agricultural Chemists for the estimation of sugar in milk, which was originally described by me in the American Chemical Journal, vol. 6, No. 5. The author is of the opinion that acetic acid and subacetate of lead can be used with equal advantage with mercuric nitrate and nitric acid for preparing the milk for polarization. He, however, prefers the use of the mercury salt. He calls attention to the correction which should be made in the reading of the polariscope for the volume of the precipitate. According to Vieth this correction in the method adopted by the association is too small. The volume of the precipitate is equal to the combined volumes of the precipitated casein and fat carried down with it. The method employed by Veith for making the correction for the volume of the precipitate is illustrated by the following example:

Given a milk with a specific gravity of 1.0325 which contains 3.72 of fat. Add to 50 cubic centimeters of this milk 1½ cubic centimeters of the mercuric nitrate solution. The polarization of the filtrate shows 5.1 milk sugar. The necessary calculations for the correction of this number are as follows:

$$93 : 100 :: 372 : X \text{ whence } X = 4$$

the volume occupied by the fat.

$$103.25 : 96 :: 51 : X \text{ whence } X = 4.74.$$

The number 96 in the above formula represents 100 cubic centimeters of the liquid less 4 cubic centimeters, the volume of the fat. The number 4.74 represents the percentage of crystallized milk sugar. The percentage of the anhydrous milk sugar would therefore be 4.5.—(Abstract from the Analyst, vol. 13, No. 144, p. 63.)

SHORT'S METHOD OF DETERMINING FAT IN MILK.

[Bul. Wis. Ag. Ex. Station, No. 16.]

The process depends on the following facts: That when a mixture of milk and a strong alkali is heated to the temperature of boiling water for a sufficient time the fat of the milk unites with the alkali and forms a soap which is dissolved in the hot liquid; at the same time the casein and albumen are disintegrated and become much more easily soluble. After the heating has continued for about two hours the mixture of milk and alkali becomes homogeneous and of a dark brown color. On the addition of an acid the soap is decomposed, the fatty acids are set free, and rise to the surface, while the albumen, casein, etc., are first precipitated and then dissolved. The insoluble fatty acids thus obtained constitute very nearly 87 per cent. of the total fat of the milk.

APPARATUS.

The process requires the following apparatus:

(1) Tubes made of soft lead glass about one-sixteenth inch thick. The lower part of the tube is about 5 inches long and fifteen-sixteenths of an inch in diameter. The upper part of the tube 5 inches long and one-fourth inch inside diameter.

(2) Three pipettes, one holding when filled up to the mark on the neck 20 cubic centimeters (about two-thirds of an ounce), this being the exact amount of milk to be taken for analysis; the other two pipettes, holding 10 cubic centimeters each, for measuring the alkali and acid used.

(3) A scale, divided in millimeters, for measuring the column of fat when the analysis is finished. The one used by the writer is a folding boxwood rule, but any rule divided in millimeters will answer the purpose.

(4) A water bath made of sheet copper. It is provided with a rack to hold the tubes while being heated; also a feed and overflow to keep the water in the bath at a constant level.

(5) A wash bottle to hold hot water.

SOLUTIONS REQUIRED.

The solutions required for the process are as follows:

No. 1.—8.75 ounces (250 grams) caustic soda and 10.7 ounces (300 grams) caustic potash dissolved in 4 pounds (1,809 grams) water. Use 10 cubic centimeters for each analysis.

No. 2.—Equal parts of commercial sulphuric and acetic acids. The acetic acid should be of 1.047 specific gravity. Use 10 cubic centimeters of the mixed acids for each analysis.

DIRECTIONS FOR ANALYSIS.

Taking samples.—Mix the milk thoroughly by pouring from one vessel to another, avoiding as much as possible the formation of air bubbles; warming the milk to 80 or 90 degrees Fahrenheit will prevent frothing to a large extent. After mixing, allow the milk to stand one or two minutes, to permit the air bubbles to escape before taking samples. Fill the 20 cubic centimeter pipette by placing the lower end in the milk and sucking until the milk rises in the tube above the mark on the side. Place the finger quickly on the top of the tube and allow the milk to run out slowly until it falls to the mark on the side of the tube; then let the contents of the pipette run into one of the analytical tubes, blowing out the last few drops.

Adding the alkali.—Fill one of the 10 cubic centimeter pipettes to the mark on the side with alkali, and allow the solution to flow into the milk just measured. Place the finger on the top of the tube and shake the tube until the milk and alkali are well mixed. A rubber cot on the finger will protect it from the action of the alkali. Treat all samples in the same way. Place the tubes in the rack, set the rack and tubes in the water bath, and heat the bath until the water boils; continue boiling for two hours, or until the contents of the tube become homogeneous and of a dark brown color similar to that of sorghum molasses. After the tubes have boiled for one hour remove the rack and tubes from the bath and examine the tubes to see if the contents are well mixed. If a whitish layer of casein and fat is found floating on the surface of the liquid gently shake and roll the tubes until the contents are well mixed. Return tubes to water bath and boil one hour. The tubes are then ready for the addition of the acid.

Adding the acid.—Remove the rack with the tubes from the water and allow them to cool to about 150 degrees Fahrenheit. Then, by means of the pipette, add 10 cubic centimeters of the acid mixture to each tube, slowly, so as not to cause the contents of the tube to froth over. Mix the acid with the contents by running a small glass tube to the bottom of the mixture and blowing gently. Place the rack and tubes again in the bath and heat to boiling for one hour. Remove the tubes from the water, and then, by means of the water-bottle, fill the tubes with *hot* water to within 1 inch of the top. The fat will then rise to the top of the water. Replace the tubes in the bath and allow them to stand in the hot water, without boiling, for one hour. At the end of this time remove the tubes from the bath, one at a time, and measure while hot.

Measuring the fat.—The analyst will observe that the lines representing the upper and lower limits of the column of fat do not extend straight across the tubes, but are slightly curved. In measuring the column of fat place the rule on the tube so that the lower line will come opposite the lowest part of the curved line of the fat; then read up the scale to the division coming opposite the lowest part of the upper curved line. The number of divisions on the rule is the length of the column of fat in millimeters. The per cent. of fat in the milk is then calculated from the following formula and data:

> Amount of milk taken, 20 cubic centimeters.
> Specific gravity of milk, 1.032.
> Specific gravity of insoluble fatty acids, .914
> Per cent. of insoluble fatty acids in butter fat, 87.

From the above data we have the following formula:

$$100 \frac{a \times b \times c}{d \times e} = x$$

Where a = the length of the column of fat in millimeters.

b = the value of one linear millimeter of measured fat expressed in cubic centimeters. The value of b will vary according to the size of the tube used.

c = specific gravity of the insoluble fatty acids.

d = 20 64 grams or the volume of milk taken for analysis multiplied by its specific gravity.

e = per cent. of fat present in sample of milk taken for analysis.

Substituting the figures obtained by an actual analysis the formula would be—

$$100 \frac{30 \times .027 \times .914}{20.64 \times .87} = x = 4.12$$

per cent. of fat in sample of milk analyzed.

(Note.—The neck of the tubes can be graduated in cubic centimeters, removing the necessity of the troublesome measurements mentioned above.—H. W. W.)

If the process has been conducted according to the above directions the column of fat will be free from impurities and the line of separation between the water and fat will be perfectly clear. On first rising to the surface the fat is slightly turbid owing to the presence of a small quantity of water. Although this will make no appreciable difference in the measurement of the fat it may, if desired, be obtained perfectly clear by removing the tubes from the bath and allowing them to cool slowly. The crystallization of the fat causes the finely divided water, which is distributed through the fat, to collect in drops, which sink to the bottom when the tubes are again heated, leaving the fat perfectly clear.

The analyst may fail to obtain correct results from the following causes: Either the column of fat may contain flecks of undecomposed casein, which would increase the volume of fat, thereby giving too high a per cent., or a small quantity of butter fat may remain unsaponified, which will also give too high results. These errors are both caused by insufficient heating of the milk with the alkali, and may be easily obviated by taking care to heat the mixture of milk and alkali for two hours at least. If not pressed for time it is better to heat two and one-half hours, and thereby remove all risks of the above errors. If milk containing more than 6 per cent. is to be tested the mixture of milk and alkali should be heated at least three hours. In such case it would be better, perhaps, to take 10 cubic centimeters in place of the usual amount.

Before adding acid the tubes and contents must be allowed to cool to 150° Fahrenheit at least. If added at a higher temperature the contact of the strong acid with the hot alkali solution will generate sufficient heat to cause the contents of the tube to boil with explosive violence, throwing out the contents of the tube and spoiling the analysis. If after the addition of hot water the tubes are allowed to stand in boiling water, small bubbles of gas are given off by the continued action of the acid on the casein. These bubbles rise through the column of fat, rendering it turbid, and causing difficulty in measuring. The bottles containing the solutions of acid and alkali should be kept corked when not in use. If the acid bottle be left open the acetic acid will evaporate and the acid will not dissolve the casein. The alkali bottle should be kept closed to prevent absorption of carbonic acid and consequent weakening of the solution.

(NOTE.—The method of Short has been compared in my laboratory with the anhydrous copper method of Piggott & Morse. A mean of 12 determinations gave by the anhydrous copper method 3.20 per cent. of fat, and by the Short method 3.18 fat. The same samples treated by the Adams method gave results considerably higher. The process of Short has also been improved by graduating the neck of the tube into tenths of a cubic centimeter.

For use at stations where expensive chemical apparatus can not be had, the method appears to have great value.—H. W. W.)

BICARBONATE OF SODIUM IN MILK.

Proust makes an energetic protest against the addition of bicarbonate of sodium to milk for the purpose of preserving it. His principal objection to the use of the above salt is based upon its action on lactic acid, the lactate of soda being a salt injurious to children.—(Abstract from Chemisches Central-Blatt, No. 24, 1888, p. 837.)

BUTTER ANALYSIS.

ACTION OF ALCOHOL ON BUTTER FAT.

Cochran has made experiments on the solubilities of the different glycerides of butter in alcohol with reference to the quantity of volatile acid which the dissolved and undissolved portions would yield when treated by Reichert's method. As a result of the work it is seen that that portion of the glycerides dissolved by ethyl or methyl alcohol is richer in volatile acids than the undissolved portions. The iodine number of the dissolved fat is less than the undissolved fat. The melting point of the fats dissolved by alcohol is less than that of the undissolved fats.

The above facts would be of importance in the examination of butters to which it is suspected artificial butyrates may have been added.—(Abstract from the Analyst, vol. 13, p. 55.)

A QUICK METHOD OF DISTINGUISHING OLEOMARGARINE IN BUTTER.

Dubernard has proposed the following method for a qualitative examination of butter for oleomargarine. The method rests upon the observation that pure butter vigorously shaken with ammonia at a temperature of from 70° to 80° and afterwards heated to 100° produces only a small amount of foam, which does not last a long time. Margarine, on the contrary, produces a large quantity of lasting foam. In order to distinguish butter from margarine by this method, the following manipulation is pursued.

In a test tube place about 3 grams of the butter to be examined. Heat to 95° or 100° in a water bath, and then allow to cool to about 80°. Add to the melted substance 5 cubic centimeters of ammonia, shake vigorously, and place again in the water bath and heat to 95° or 100°. The ammonia is volatilized with the formation of foam. The foam in the tube rises more or less according as the butter contains a larger or smaller quantity of oleomargarine. In mixtures of pure butter with known quantities of margarine the tube can be graduated and the relative quantities of butter and margarine thus approximately determined.—(Abstract from Chemiker-Zeitung, 1888, No. 46, p. 760.)

DETECTION OF ADULTERATIONS IN BUTTER.

Bockairy proposes the following method for the detection of adulterations in butter which rest upon the different solubilities of fats in toluene :

Place in a test tube 15 cubic centimeters of pure toluene, add 15 cubic centimeters of the filtered fat and 40 cubic centimeters of strong alcohol at 18°; the toluene holding in solution the fatty matter remains at the bottom of the tube. By means of a water bath the tube is heated to 50° and thoroughly shaken. If the sample is of butter fat or a mixture of butter fat and some other kind, there is no turbidity; but if no butter be present a turbidity is at once manifested. The tube is placed in water at a temperature of 40° for half an hour; pure butter gives no turbidity at the end of that time, but if other fats are present the mixture will appear turbid.—(Abstract taken from the Chemical News, July 6, 1888, p. 11, and Bul. de la Societe Chimique de Paris, vol. 49, No. 5, p. 331.)

DIFFERENCE BETWEEN NATURAL AND ARTIFICIAL BUTTER.

C. J. van Lookeren proposes the following for a characteristic test between pure and falsified butter:

A small amount of butter is melted in a teaspoon, and a drop thereof placed in boiling water contained in a watch glass. If the butter be pure a thin film of fat is formed upon the hot water, which breaks up into numerous fat globules, which tend to collect quickly at the periphery. With oleomargarine, etc., a thin film of fat is formed in the same way, which, however, breaks up into only a few large drops, which remain distributed over the whole surface of the water. The conditions for the success of the experiment are that the water be perfectly pure and clear, and the melted butter fat very hot.—(Abstract in the Repertorium of the Chemiker-Zeitung, No. 18, 1888, p. 143, from the Milchztg, 1888, No. 17, p. 362.)

A MODIFICATION OF KOETTSTOEFER'S AND REICHERT'S PROCESSES.

Lowe has proposed the following method of determining the volatility and saponification equivalent of butter and other fats:

About 2 grams of the filtered fat are saponified with 10 cubic centimeters of normal alcoholic potash in a stoppered flask. The alcohol is then boiled off and the soap dissolved in 50 cubic centimeters of hot water. The excess of potash is then determined with semi-normal sulphuric acid solution. The saponification numbers having thus been determined enough additional sulphuric acid is run in to make 25 cubic centimeters in all. The flask is then connected to a condenser and 50 cubic centimeters distilled off, filtered, and determined in the usual way. The same flask is used throughout the whole process, one of 200 cubic centimeters capacity answering the purpose very well. The whole process does not occupy more than two hours, and can be completed often in less time.

Tyrer, criticising the process of Mr. Lowe, was of the opinion that the quantity of fat taken, 2 grams, was too small, since it gave only about one-tenth of a gram of volatile fatty acids. The alkali also dissolved some of the glass, thus introducing an error of which no account was taken. Carbonic acid, moreover, expelled by this method, increased the apparent estimate of filtered fatty acids.—(Abstract from the Journal of the Society of Chemical Industry, March, 1888, p. 185; May, 1888, p. 376.)

SAPONIFICATION WITHOUT THE USE OF ALCOHOL.

Mansfeld, in order to avoid the losses due to etherification during saponification in the presence of alcohol, has proposed to carry on the process of saponification, preparatory to the estimation of volatile acids in butter, without the use of alcohol. The method employed is as follows:

Five grams of the melted and filtered butter fat are taken; into the melted fat are allowed to run 2 cubic centimeters of a potash lye containing 100 grams caustic potash in 100 cubic centimeters of water. The flask is closed with a stopper carrying a glass tube drawn out to a capillary point; the flask is then placed in an air bath heated to about 100° and allowed to remain for two hours. At the end of that time the saponification is completed. One hundred cubic centimeters of water are now added, the flask placed in a water bath until the soap is dissolved, which is decomposed and subjected to distillation in the usual way.—(Abstract in Chemisches Central-Blatt, June 23, 1888, p. 870, from M. Z., No. 17, pp. 281-83).

WOLLNY'S CRITICISM OF THE REICHERT-MEISSL METHOD.

The method of determining volatile fatty acids devised by Reichert has, during the past year, been subjected to an extended criticism by Dr. R. Wollny, of Kiel. Dr. Wollny, as a result of 98 analytical tests, was convinced of the truth of the statement of Professor Fresenius, that the Reichert method was totally unreliable for the determination of very small quantities of butter in oleomargarine.

As a result of his analytical work Dr. Wollny was convinced that the chief source of error in the Reichert work was due to the absorption of carbonic acid. He asserts that the error which may arise from this source must amount to as much as 10 per cent. butter, and renders the results obtained by the method quite inaccurate. It being very difficult to secure an alkali free from carbonate, the author decided to use a 50 per cent. solution of caustic soda instead of potash, since in a solution of that strength the chloride, nitrate, sulphate, and carbonate of soda are quite insoluble·

An arrangement was also devised by which the solution of caustic soda could be kept and drawn off for use without risk of absorption of carbonic acid. Such a solution made and kept in this way gave for 3 cubic centimeters treated by the ordinary process of Reichert's distillation an amount of volatile acid sufficient to neutralize from .2 to .3 cubic centimeters of the deci-normal barium hydrate solution. The author's precautions to prevent absorption of CO_2 are wholly unnecessary when the saponification is carried on in closed flasks.

Dr. Wollny further discusses the errors due to the formation of butyric ethers during saponification and distillation, and also the error due to the mechanical translation of particles of insoluble fatty acids during the process of distillation, and the error due to the shape and size of the vessel in which the distillation is carried on and the time of its duration. The magnitude of these errors he states as follows:

(1) Due to absorption of carbonic acid +10 per cent. (2 and 3) Formation of butyric ether —13 per cent. (4) Cohesion of fatty acids —30 per cent. (5) Shape and size of vessel + or —5 per cent.—(M. Z., 1887, Nos. 32, 33, 34, 35, and the Analyst, 1887, Nos. 139, 140, 141, 142.)

Dr. Wollny further describes the exact methods to be employed in the examination and the results obtained, for the details of which I refer to the original papers.

As president of the butter commission of the German Dairy Association, Dr. Wollny has proposed the following method to be employed in the comparative examinations of butter for the forthcoming report of the commission (see M. Z. No. 25, 1888). In the preparation of the deci-normal barium solution each one of those taking part in the analyses has been furnished with samples of normal sulphuric acid, pure crystallized chloride of barium, and pure peroxalate of potash. In addition to this for testing the refractometer samples of olive oil, nitrobenzol and monobromnaphtalin have been sent. Following are the details given by Dr. Wollny for the preparation of the reagents and the conduct of the analysis.

METHOD PROPOSED BY DR. WOLLNY FOR THE EXAMINATION OF BUTTER FOR THE COMMISSION OF THE GERMAN DAIRY UNION.

[From the Milch Zeitung, 1888, No. 25 et seq.]

(1) PROVING THE WEIGHTS AND BURETTES.

The weights which are to be used must be carefully compared and the burettes calibrated.

(2) ESTIMATION OF THE STRENGTH OF NORMAL SULPHURIC ACID.

A clean dry dropping bottle of about 60 grams content, the tip of which is touched with vaseline on the outside, is carefully weighed; 35 or 40 grams of normal sulphuric acid are then placed in it and it is again weighed. Afterwards five portions in duplicate of about 3 grams are weighed in a beaker glass of 300 cubic centimeters capacity. The drop bottle being weighed after each portion, the exact weight of the normal sulphuric acid is determined. One hundred cubic centimeters of recently boiled distilled water are added to each of the portions. One hundred cubic centimeters of this distilled water must not require more than 2 drops of barium solution to give color after 1 cubic centimeter of the phenolphtalein solution has been added. The samples are heated in the water bath to the boiling point and to each portion as many cubic centimeters of dilute barium chloride solution added that for each gram of the normal sulphuric acid there are present 10 cubic centimeters of the BaCl₂. The barium chloride solution contains 15 grams of the salt in 1 liter of distilled water. Afterwards the beakers are covered with watch glasses and allowed to stand for fifteen minutes over the water bath. The precipitates are to be collected upon ash free filters of 9 centimeters diameter and washed with hot distilled water until the chlorine action disappears. After drying and incineration the weight obtained is to be multiplied by the factor .34331. The five duplicate portions, after the addition of 1 cubic centimeter of phenolphtalein solution, are treated with barium solution until the red color appears. The numbers obtained are to be calculated to 100 grams of the normal sulphuric acid.

(3) ESTIMATION OF SATURATION OF THE NORMAL SULPHURIC ACID BY THE BARIUM SOLUTION.

From the results obtained by the above analyses the exact strength of the deci-normal barium solution is to be determined.

(4) TITRATION OF BARIUM SOLUTION WITH POTASSIUM PEROXALATE.

One gram of the peroxalate of potassium, after drying for twelve hours in a desiccator over sulphuric acid, is placed in a weighing bottle similar to the one described above.

Sixty grams of hot distilled water are then added, and after the salt has been dissolved and the solution cooled the weighing bottle is again weighed. About 12 grams of the solution are now run into each of five beaker glasses, and the exact amount in each one determined by reweighing the weighing bottle. Ninety cubic centimeters of recently boiled distilled water are now added, together with 1 cubic centimeter of phenol phtalein solution. The barium solution is now added until the red color, which at first disappears, remains for at least five minutes. The results are calculated to the strength of the barium solution according to the formula—

$$T = \frac{10000 \times p\ (b-t)}{84.5\ n\ (l-t)}$$

in which

$t =$ the weight of the weighing bottle empty.
$b =$ the weight of the weighing bottle + the peroxalate.
$l =$ the weight of the weighing bottle + the peroxalate solution.
$p =$ the weight of the portion of the solution employed.
$n =$ the number of cubic centimeters of barium solution used.

(5) ESTIMATION OF THE VOLATILE ACIDS AFTER SAPONIFICATION WITH THE AID OF ALCOHOL.

Five grams of the butter fat are weighed into an Erlenmeyer flask; 10 cubic centimeters of alcohol at 96 per cent. and 2 cubic centimeters of concentrated soda lye at 50 per cent., which has been preserved in an atmosphere free of carbonic acid, are added. The flask, furnished with a reflux condenser, is heated, with occasional shaking, in a boiling water bath for one-quarter of an hour. The alcohol is then distilled off by allowing the flask to remain for three-quarters of an hour in a boiling water bath. One hundred cubic centimeters of recently boiled distilled water are then added and allowed to remain in the water bath until the soap is dissolved. The soap solution is then immediately decomposed with 40 cubic centimeters of dilute sulphuric acid (25 cubic centimeters sulphuric acid to 1 liter), and the flask immediately connected with the condenser. This connection is made by means of a 7 millimeters diameter glass tube which, 1 centimeter above the cork, is blown into a bulb 2 centimeters in diameter; the glass tube is now carried obliquely upwards about 6 centimeters and then bent obliquely downward; it is connected with the condenser by a not too short rubber tube. The flask is now warmed by a small flame until the insoluble acids are melted to a clear transparent liquid. The flame is now turned on with such strength that within half an hour exactly 110 cubic centimeters are distilled off. One hundred cubic centimeters of the distillate are now filtered off, placed in a beaker glass, 1 cubic centimeter of phenolphtalein solution added and titrated with barium solution; when the red color is shown the contents of the beaker glass are poured back into the measuring glass in which the 100 cubic centimeters was measured, again poured back into the beaker, and again titrated with the barium solution until the red color becomes permanent. The distillation should take place in as nearly thirty minutes as possible.

(6) ESTIMATION OF THE VOLATILE ACIDS AFTER SAPONIFICATION WITHOUT ALCOHOL.

Five grams of the butter fat are saponified with 2 cubic centimeters of concentrated potash lye preserved from contact with carbonic acid. The lye solution is made by dissolving 100 grams of the potash in 58 grams of water. By gentle rotation of the flask the lye and fat are intimately mixed together; the flask is then placed in a vertical position over a boiling water bath until the mixture becomes solid. It is then placed obliquely in the water bath. After it has remained here for two hours it is taken out and the remainder of the process continued as above.

7717—No. 19——2

(7) ESTIMATION OF THE MEAN MOLECULAR WEIGHT OF THE VOLATILE FAT ACIDS.

The soap solution obtained by the titration of barium hydrate in number 6 is placed in a weighed platinum dish evaporated to dryness in the water bath, dried in a drying oven for two hours and again weighed.

(8) ESTIMATION OF VOLATILE FAT ACIDS BY DISTILLATION FROM MAGNESIUM SALTS.

Eight grams of butter fat are placed in an Erlenmyer flask, treated with 3 cubic centimeters of concentrated potash lye and 15 cubic centimeters of alcohol as in number 5, and the alcohol then distilled off. The soap, dissolved in 100 cubic centimeters of water, is washed with 250 cubic centimeters of water in a measuring flask of 500 cubic centimeters capacity and cooled to the temperature of the room. Then by means of a pipette, with constant shaking, 50 cubic centimeters of magnesium sulphate solution, 15 per cent. strength, are added, the flask filled up to the mark with water and vigorously shaken. The mixture is now placed upon an 15 centimeter filter and 300 cubic centimeters filtered into a measuring flask. The filtrate is brought into a flask holding about 500 cubic centimeters, which on the one side is joined with a steam generator and on the other with a condenser. Two cubic centimeters of strong sulphuric acid together with two pieces of pumice stone are added. Two hundred and fifty to 275 cubic centimeters are now distilled off into a measuring flask of 500 cubic centimeters capacity, the distillate being filtered into the flask through a moistened filter. The gas flame under the distillation flask is then turned down low and the distillation continued in a current of steam until the 500 cubic centimeters flask is full. After the filter which has been used has been washed with a little water the distillate is placed in a large flask of 1 liter capacity and 2 cubic centimeters of phenolphtalein added and titrated.

(9) ESTIMATION OF THE VOLATILE FATTY ACIDS FROM COPPER SALTS.

The estimation is carried on exactly as in No. 8 substituting 50 cubic centimeters of copper sulphate solution for the 50 cubic centimeters of magnesium salts. In the present case the distillate may be collected directly in the one-half-liter flask without filtration, since by precipitation with copper sulphate solution no insoluble fatty acids pass over during distillation.

(10) ESTIMATION OF VOLATILE FATTY ACIDS BY DISTILLATION IN A CURRENT OF STEAM, ACCORDING TO GOLDMANN.

Five grams of butter fat are weighed in a long-necked flask of about 300 cubic centimeters capacity. Ten cubic centimeters alcohol and 2 cubic centimeters potash lye are added and the flask connected, on the one hand, with a steam generator, and on the other with a condenser. The contents of the flask are now warmed with a small flame for twenty minutes the condenser being directed obliquely downward and, in a measuring cylinder, 6 cubic centimeters distilled off. The steam is then directed into the flask and 50 cubic centimeters more collected. The soap is now decomposed with 5 cubic centimeters of sulphuric acid (20 cubic centimeters strong acid diluted to 100 cubic centimeters) the separated fat acids heated with a small flame until they form a clear solution and in a current of steam the volatile acids distilled off into a 500 cubic centimeter flask. The size of the flame is so regulated that the contents of the distillation flask remain constant at about 30 to 40 cubic centimeters, and the distillation is continued until each succeeding 500 cubic centimeters, after the addition of 2 cubic centimeters phenolphtalein solution, require not more than 2 cubic centimeters of the deci-normal barium solution to produce the red color.

(12) ESTIMATION OF THE FATTY ACIDS SOLUBLE IN 10 PER CENT. ALCOHOL.

(a) *Roses method.*—12.5 grams of butter fat are weighed into a graduated flask of 500 cubic centimeters capacity, 50 cubic centimeters of alcoholic potash lye (about 112 grams of potassium hydrate dissolved in absolute alcohol and made up to 1 liter), are added and gently shaken. After five minutes as slight an excess of semi-normal sulphuric acid as possible is added. After two minutes dilute with water to 490 cubic centimeters; to destroy the foam, add 5 cubic centimeters more absolute alcohol, fill up to the mark with water and then add as much more water as will correspond to the amount of fat taken. The flask is now shaken, contents filtered through a dry filter and 250 cubic centimeters of the filtrate titrated with deci-normal potash lye.

(b) *Modification of the above process by Dr. Wollny.*—Weigh 6.25 grams of the fat in an Erlenmeyer flask of 300 cubic centimeters capacity. Add 25 cubic centimeters of potash lye, made with absolute alcohol as described above, and then diluted with absolute alcohol until not more than 74 cubic centimeters and not less than 73 cubic centimeters of semi-normal sulphuric acid are required to neutralize it. The flask is then closed with a stopper which carries a glass tube 10 millimeters wide and 50 centimeters long; heat with a very small flame until the contents are near the boiling point for about twenty minutes and until the smell of the butter ether has entirely disappeared. The flask is then cooled in water and 150 cubic centimeters of recently-boiled distilled water added. When the soap is dissolved and the whole mass cooled to about 15°, 75 cubic centimeters semi-normal sulphuric acid are added. The flask is then closed with a rubber stopper and shaken vigorously for one minute. Filter through a strong dry-folded filter into a graduated flask holding 200 cubic centimeters, pour the 200 cubic centimeters into a large beaker glass, and titrate with barium solution.

(13) ESTIMATION OF INSOLUBLE FAT ACIDS.

Treat 3 to 4 grams of the butter fat with 2 cubic centimeters concentrated soda lye and 10 cubic centimeters alcohol in a porcelain dish on the water bath and evaporate the soap to dryness. Dissolve the soap with 100 cubic centimeters of hot water, add 5 cubic centimeters concentrated sulphuric acid, and heat the separated fatty acids for one hour on the water bath. Pour the contents of the dish on a filter of 11 centimeters diameter, made of the best thick Swedish filter paper previously dried and weighed. Upon this filter the acids are washed with 1.5 liters of boiling water. The filter with the insoluble fatty acids, after they have solidified, is placed in a weighed beaker glass, dried for two hours in an air bath, and weighed.

(16) ESTIMATION OF THE FREE FAT ACIDS.

Ten grams of the butter fat are placed in a beaker glass with 20 cubic centimeters of ether and 10 cubic centimeters of alcohol and 1 cubic centimeter phenolphtalein solution. The free acids are then titrated with deci-normal alcoholic potash.

(11, 14, 15, 17, 18, 19) COMBINED ESTIMATION OF THE SAPONIFICATION NUMBER, THE VOLATILE, SOLUBLE, AND INSOLUBLE FAT ACIDS.

(a) *Saponification number.*—Five grams of the fat are weighed into an Erlenmeyer flask and 25 cubic centimeters of alcoholic potash added. (Twenty-five cubic centimeters alcoholic potash should saturate from 45 to 50 cubic centimeters semi-normal sulphuric acid.) The mixture should be warmed as in No. 5 for a full hour with the reflux condenser. Meanwhile the same quantity of alcoholic potash should be exactly titrated with the semi-normal sulphuric acid, and later being titrated by the barium solution used and the number obtained noted. The alcoholic soap solution is taken from the waterbath, the solid parts adhering to the walls of the flask dissolved by gentle shaking, and, after the addition of 1 cubic centimeter phenolphtalein solution, titrated with semi-normal sulphuric acid. The number obtained is subtracted from the number obtained in the blank experiment and calculated back to the corresponding number of cubic centimeter's barium solution.

(b) *Volatile fat acids.*—The alcohol is driven off from the exactly neutralized alcoholic soap solution obtained in the foregoing experiment, and then an additional amount of sulphuric acid added, amounting to 2 cubic centimeters more than necessary to neutralize the original potash lye employed. The mixture is then diluted with 140 cubic centimeters of recently-boiled distilled water and two pieces of pumice stone added, and, as in No. 5, 110 cubic centimeters distilled off, of which 100 cubic centimeters, after filtration, is titrated.

(c) *Soluble fat acids.*—After the end of the foregoing experiment the distillation flask is removed from the distillation tube and replaced by one filled with distilled water and placed as a receptacle under the condenser. The condensation water is now allowed to flow from the condenser and the condensed insoluble fatty acids driven over into the original distillation flask by a current of steam. In like manner the insoluble acids collected upon the filter are washed into the original distillation flask. After the fatty acids are solidified the whole of them are brought into a cylinder 3.5 centimeters wide and 15 centimeters long, which ends in a 6-millimeter wide tube carrying a goose-necked glass tube from its side; and on the lower end is closed with a rubber tube and pinch cock. The lower contraction of the cylinder is stopped with a plug of glass wool which forms a thick filter but allows hot water to run through easily. Under the goose neck of this wash vessel, which is held in a vertical position by a retort holder, is placed a graduated flask of 500 cubic centimeters capacity into which the excess of water flows. When the acids have been brought from, the distillation flask, together with the pieces of pumice stone into the wash flask, any residue is washed out with the washing bottle with boiling water. The above must be so conducted that no drops of the fatty acid pass through the glass wool and that no fatty acid remains on the upper walls of the washing flask. The washing flask is furnished with a gum stopper carrying two holes. Through the one passes a tube for the inlet of steam; through the other a tube at least 8 millimeters in diameter, cut off obliquely at the lower end and divided above the stopper into two branches. By means of one of these branches it is connected with a perpendicular condenser, at least 8 millimeters wide, and through the other it is connected by means of a rubber tube and pinch cock with a hot-water reservoir by means of a U-tube reaching to the bottom thereof. Through the steam tube, which is somewhat bent at the end, so that the steam is directed obliquely downwards and outwards against the sides of the washing bottle, a current of steam is conducted which stirs up in a lively manner the surface of the fatty acids, which swim upon the top of the water, while the condensed water drops back out of the reflux condenser in a boiling-hot condition and the excess flows through the goose neck into the measuring flask. The wash water collected in the 500 cubic centimeter flask is titrated with barium solution, with the addition of 2 cubic centimeters phenolphtalein solution. This operation is repeated three times; in all, 1.5 liters of wash water being collected. The total amount of barium solution required in the three operations is collected in one number.

(d) *Insoluble fatty acids.*—After the end of the foregoing experiment the wash flask is cooled until the fatty acids are solidified. The pinch cock closing the lower opening of the wash flask is opened, the water allowed to flow off, and a current of air passed through the apparatus by means of an aspirator after the opening of the goose neck is lightly stoppered and the steam generator removed. The wash flask together with the fatty acids and the glass wool plug are completely dried in this manner after, at most, half an hour. The steam generator is now replaced with a flask which contains ether free of water, alcohol, and acids. The ether is brought to the boiling point by means of warm water and its vapor condensing in the washing flask dissolves the fatty acids and carries them into the weighed flask of about 100 cubic centimeters capacity placed below to receive them. When about 40 or 50 cubic centimeters of ether have distilled over, the operation is broken, the outer end of the wash flask washed with ether and the weighed flask below is placed in a warm place in order to drive off the ether. The weighed flask containing the acids is now warmed in the water bath and placed under the receiver of an air pump. After the acids are completely freed from ether by this method the flask is again weighed and the weight of the acids noted.

(e) *Mean molecular weight of the insoluble fat acids.*—The fatty acids obtained in the foregoing experiment are melted at a low temperature and from .8 to 1 gram weighed in a beaker glass, dissolved in 50 cubic centimeters of alcohol, 1 cubic centimeter phenolphtalein solution added and titrated with barium solution.

(20) DETERMINATION OF THE IODINE NUMBER.

Weigh out from .8 to 1 gram butter fat in a flask of about 200 cubic centimeters capacity furnished with a glass stopper; dissolve in 10 cubic centimeters chloroform, and add 20 cubic centimeters iodine solution. In case the liquid is not clear, more chloroform must be added. If the iodine color should rapidly disappear add from 5 to 10 cubic centimeters more. The amount of iodine added should be sufficient to secure strong colorization after standing two hours. Add 10 to 15 cubic centimeters iodide of potassium mixture and 150 cubic centimeters of water; titrate with thiosulphate of sodium solution until the iodine has almost disappeared. Add then a little starch solution, and continue titration until the blue color has disappeared.

(21) ESTIMATION OF REFRACTIVE INDEX.

Set the large Abbe refractometer with water at 18° so that the index reads 1.3330. The instrument should then be placed in a room which is kept at a temperature of about 25°. The estimation of the refractive index of butter fats can not be made below 25° on account of the solidification of the fat. The instrument is also to be tested at 20° with olive oil, nitrobenzol and monobromnaphtalin.

(25) DETERMINATION OF THE SPECIFIC GRAVITY AT ZERO OR 15°.

A platinum crucible is supplied with a platinum wire handle so that it can be suspended from the hook of the balance. It is then weighed empty and afterwards its specific gravity taken in water at zero. The water is kept at zero by being placed in a small beaker glass, which in turn is put in a larger beaker closed and surrounded with finely pounded ice. The crucible is then dried and 15 grams of melted butter fat placed therein. The butter fat is allowed to solidify slowly at a temperature not below 15°. The crucible with its contents is then weighed in the air and placed in distilled ice-cold water for an hour. The whole is then weighed in water at zero as before. From these results the specific gravity of the fat at zero is determined. In the same way determine the specific gravity at 15°.

(26) ESTIMATION OF THE SPECIFIC GRAVITY IN BOILING WATER BY MEANS OF THE PYK-NOMETER.

A small flask of 50 grams, the neck of which is drawn out to a narrow tube and marked at the narrow place, is weighed empty. It is then filled with distilled water to the mark and placed for an hour in finely pounded ice. The contraction of the water is restored by the addition of ice-cold water until the mark is reached; it is then dried, allowed to come to the temperature of the room, and weighed. The same flask is then filled to the mark with boiling water, and, after coming to the temperature of the room, weighed. The flask is then emptied, dried, filled with melted butter fat, kept in a boiling-water bath for an hour, the butter fat filling to the mark, afterwards cooled to the temperature of the room, and weighed.

(27) ESTIMATION OF THE SPECIFIC GRAVITY WITH MOHR'S BALANCE.

This is determined in the usual manner by the use of a specific gravity bob, the thermometer of which is arranged so as to show a temperature of 100°.

(28) ESTIMATION OF THE SPECIFIC GRAVITY WITH THE AREOMETER IN BOILING WATER.

This is determined by means of spindles furnished by the firm of G. C. Gerhard, in Bonn. These spindles are graduated especially for specific gravity at a temperature of 100°.

(29) ESTIMATION OF THE MELTING POINT.

This is determined in straight or U-formed capillary tubes. They are filled with butter fat at 100° and placed in ice-water for a quarter of an hour. The capillary tube is then placed, with a thermometer, in a beaker glass with water at 20° and slowly warmed with a small flame.

(30) ESTIMATION OF THE SOLIDIFYING POINT.

Place about 100 cubic centimeters of the melted fat in a test tube at 40°. Stir with a thermometer until the fat begins to solidify, the test tube meanwhile being placed in a large vessel containing water at 20°. The temperature will rise slightly when the solidification begins, and the highest point reached is entered as the solidification point.

NOTE.—Dr. Wollny has proposed in the above scheme an exhaustive study of butter fat. Many points appear to be superfluous and others elaborated into unnecessary detail. The method of determining insoluble fatty acids appears to be more objectionable than any other part of the scheme. The method of making the melting point first described in the Journal of Analytical Chemistry (vol. 1, No. 1, p. 39) appears to have escaped the notice of the German commission. I also note with regret the apparent ignorance of the existence of the Gooch crucible existing in Germany.

Points to be especially commended in these investigations are, sending samples and standard re-agents, blanks for entering the analytical results, and minute printed directions for conducting the manipulation.

The numbers prefixed to the several paragraphs correspond to the number of the column in the blank in which the results are to be entered.—H. W. W.

CRITICISM OF WOLLNY'S METHODS.

Goldmann criticises Wollny's original method chiefly because it gives only a part of the volatile acid present and takes no account of the whole. It has never been claimed, however, for the Reichert process that it gave anything more than a portion of the volatile acid present, but the percentage of the total volatile acids obtained is approximately constant, and therefore the results are comparable among themselves.

Goldmann proposes to remove the last traces of alcohol after saponification by distillation in a current of steam. Before the distillation of any part of the alcohol, however, the solution is heated with a reflux condenser for half an hour so that any ethers which have been formed may be recombined by the excess of alkali present. The apparatus for the removal of the last traces of alcohol is illustrated in No. 14 of the Chemiker-Zeitung for 1888, page 216.

Goldmann's paper is so long that for all details I can only refer to the original. The method finally adopted by him is as follows: 5 grams of the melted butter fat are weighed in a flask of 300 cubic centimeters capacity, the neck of which is 12 centimeters long and 2 centimeters wide. Add 10 cubic centimeters alcohol, 96 per cent., and 2 cubic centimeters of a 50 per cent. aqueous soda lye made and preserved out of contact with carbonic acid. The flask, furnished with a reflux condenser, is heated for twenty minutes with a small flame. The condenser is turned to an angle of 55° and 6 cubic centimeters of the alcohol distilled off. The flame is then extinguished, the apparatus connected with the flask delivering steam and the whole of the alcohol driven off by a slow current of steam.

Full directions for the conduct of this part of the work are given by Goldmann.

After the alcohol is driven off the soap is decomposed by the addition of 5 cubic centimeters of sulphuric acid containing 200 cubic centimeters of strong acid to the liter. The volatile acids are distilled off in a current of steam and 600 cubic centimeters of distillate taken which is titrated with one-tenth normal solution of barium hydrate.—(Abstract from Chemiker-Zeitung 1888, No. 12, p. 183; No. 14, p. 216; No. 20, p. 317.)

APPLICATION OF THE REICHERT-MEISSL-WOLLNY METHOD TO ITALIAN BUTTERS.

Besana has made an extended examination of 114 samples of butter, and from the study of them draws the following conclusions:

One hundred and fourteen samples of butter examined belong to 30 provinces. These show a melting-point from a minimum of 33° C. to a maximum of 37.7°, not including one sample which seems to be essentially different in constitution. No positive relation has been established between the point of fusion and the quantity of volatile acids; therefore the knowledge of the first offers no special utility.

The percentage of volatile acids, or the quantity of the same expressed in cubic centimeters of deci-normal alkaline solutions, shows a variation from a minimum of 21.83 to a maximum of 30.19. The classification of the 114 samples of butter, according to their quality or condition, would be as follows:

Volatile acids from	No. samples.
29 to 30. 19	23
28 to 29	25
27 to 28	49
26 to 27	9
25 to 26	3
24 to 25	2
23 to 24	1
21. 80 to 22	2
	114

The variation in the quantity of the volatile acids, in the different butters, already recognized by various experimenters in Germany and Sweden, is confirmed by these researches also in the Italian butters. Even the butters of the same dairy may present quite different qualities, although made within a very few days of each other. Upon the cause of this variability I have not ventured, up to the present time, to present any theory, and I have not even sufficient data for the support or contradiction of the theories of others. I hold the causes to be biological, and I would like the researches tending to their discovery to be made by following very closely the chemistry of nutrition and of the lacteal secretions or the transformation of aliment into milk, keeping note of the quantity and quality of the food, the stage of lactation, the condition of health in the animal, etc.; but such researches, in my opinion, should be followed upon individual cases, instead of upon a group of milch cows.— (Prof. Carlo Besana, extract from the journal "Le Stazioni Sperimentali Agrarie Italiane," 1888, Vol. XIV.)

The report being open to discussion, inquiries were made as to the cost of the lactocrite. Professor Myers stated that when adapted to the ordinary De Laval separator it cost $125, and could be obtained from the De Laval separator Company, 221 Dock street, Philadelphia, Pa. If it were necessary to purchase the separator frame in addition, the cost would probably reach nearly $250.

In regard to the cost of the lactocrite, I would say that I recently secured one for the West Virginia Experiment Station for $125 from the De Laval Separator Company, of Philadelphia.

It is arranged in such a way that it can be set in the place of the drum of the separator and run by the power that runs the latter. The construction is very simple, and for the rapid testing of milk it appears to be the most satisfactory method yet devised.

There is likewise an apparatus provided by the same company which is intended to mix artificial fats with the skimmed milk. It is arranged so that the liquid fat and the skimmed milk are violently whipped together, the fat being delivered in a finally, divided condition.

I have been suspicious for a considerable time that much of the cream now in the market is cream manufactured in this or some similar manner, and, as we have ordered one of these machines for the experiment station, we propose in the near future to see to what extent we can mix fat of cotton, pig, or cow with skimmed milk to supply the gastronomic needs of our national capital.

Mr. Richardson then made the following remarks on the milks found on sale in Washington, D. C.:

THE MILK OF WASHINGTON NOT NORMAL.

During the past year complaint and observation having led me, as Chemist of the District, to believe that the milks of the city were not of that quality which are found in other cities, where a thorough inspection and milk control are carried on, I have made an examination of eleven samples from the principal dairies and milk routes of the town, and the results are presented here with the view of showing the value of the table of normal relation of fat to specific gravity and solids as given by Hehner and Richmond, for detecting milks which have been tampered with in some one of the several ways now in common vogue.

The method of determining specific gravity was that of the piknometer, controlled by the spindle, at the standard temperature 15° C. Solids were determined on asbestus fiber in nickel dishes at 100° C. for one hour after drying on the steam-bath. Fat was determined with previously, extracted paper coils and Squibbs ether.

The results were as follows:

No.	Dairies and milk routes.	Specific gravity.	Solids.	Solids not fat.	Fat.	Normal fat.
508	First street, between Pennsylvania avenue and B street. northwest, small grocery	1.0258	13.54	8.93	4.61	5.36
515	Wise's dairy.................................	1.0334	11.38	7.67	3.71	2.38
516	Fairfax dairy	1.0341	11.68	7.88	3.80	2.59
528	Alpha dairy, 811 North Capitol street........	1.0508	12.45	8.48	3.97	3.95
529	Small grocery, Four-and-a-half street, between Pennsylvania avenue and C street .	1.0297	11.53	8.14	3.39	3.44
530	Fort Baker dairy, Third street and Indiana avenue............................	1.0320	11.79	8.73	3.06	3.10
531	Russell's dairy, Third and C streets	1.0296	10.71	7.43	3.28	2.74
538	Excelsior dairy, 1757 Pennsylvania avenue.	1.0292	11.46	8.08	3.38	3.47
539	Thompson's dairy, 511 Four-and-a-half street, southwest	1.0311	12.47	8.59	3.88	3.95
540	Floral Hill dairy...........................	1.0270	9.87	7.33	2.54	2.49
541	F. K. Ward's wagon........................	1.0343	11.39	8.31	3.08	2.31
	Average		11.66	8.15	3.51	

The milks were collected between March 23 and 30. Of the eleven samples, but three were at all near the standard which would allow their sale in Massachusetts, New York, or New Jersey, and of these three but two were normal in composition. Of the entire eleven only six were normal, according to the tables, these showing a very close agreement with theory, while the other five were so extremely discordant as not to permit of errors of analysis. The normal milks agreed remarkably well, as may be seen from the following figures:

Fat found by analysis.	Fat calculated for normal milk.	Difference.
3.97	3.95	+.02
3.39	3.44	−.05
3.06	3.10	−.04
3.38	3.47	−.09
3.88	3.95	−.07
2.54	2.49	+.05

These figures certainly show the tables of Hehner & Richmond to be equally applicable to American and continental milks.

The differences in the remainder of the milks were as follows:

Fat found by analysis.	Fat calculated for normal milk.	Difference.
4.61	5.36	− .75
3.71	2.38	+1.33
3.80	2.59	+1.21
3.25	2.74	+ .54
3.08	2.31	+ .77

In all but one case an excess of fat was found, which leads to the belief that oleo oil or similar substance had been churned into the milk to enrich a poor or skim milk, the difference being, it seems to me, altogether too large and striking, after the coincidences found in the other samples, to make it readily supposed that the samples were normal in character.

The extremely poor character of the milks of this city point to the necessity of milk inspection, even if this is due to low-grade cattle and poor feed alone; and it would seem that the tables given by Dr. Wiley may prove of great value in pointing out suspicious milks for further examination.

Dr. Wiley spoke of Hehner & Richmond's recommending the rejection of all analyses not agreeing with their tables as erroneous, and said that this was certainly very dogmatic, as on adulterated milks such agreement is impossible.

Mr. Richards spoke of Swedish machines for taking out cream and putting back cotton-seed oil, illustrated in *London Engineering*.

Professor Meyers spoke of the methods of emulsifying oil and skim-milk and its frequency. Cotton-seed oil can be switched in for calf feed, but he did not think artificial cream can be made so. Oleo oil is probably used.

The secretary then read the following communication from Professor G. E. Patrick, of Iowa:

NOTES ON THE METHOD FOR SOLUBLE AND INSOLUBLE FAT ACIDS.

In the determination of soluble and insoluble fat acids I have recently found it convenient, as well as economical of time and material, to change slightly the mode of procedure as laid down in the method (originally Muter's) adopted by this association at its last meeting (1887) and published in Bulletin No. 16 from the Department of Agriculture, Chemical Division, pages 70 and 71.

The purpose of the modification is to avoid the use of a separate saponifying bottle, and the consequent transfer of the soap solution to an Erlenmeyer flask, and evaporation of the alcohol added in this transfer.

This end is accomplished by saponifying directly in the Erlenmeyer, the latter (with stopper tied down) being inclosed in a "conversion" jar with tight cover, and containing absolute alcohol (50 cubic centimeters), the vapor of which gives an outside pressure on the Erlenmeyer equal to that within. The jar is made with vertical sides, and is only a little larger than the flask it is to inclose; its cover is held down by a screw-clamp in the usual way, a disc or washer of rubber making the union tight.

Breakage by contact of the flask with the walls of the jar while agitating is prevented by a ring of corks strung upon a wire. Thus arranged, *with a tight jar*, there is no danger of breaking the Erlenmeyer if the work is executed with ordinary care. Fifty cubic centimeters of absolute alcohol in the jar will suffice for many analyses.

Two styles of jar have been tried, one of heavy glass, iron bound (made by Eimer Amend), the other of heavy sheet copper, made (excepting the clamp) by an ordinary tinsmith; the cover of each is a thick ground-glass plate, with a metal disk to receive the pressure of the screw. The glass one, allowing a view of the interior during saponification, is the more convenient.

The saponification is conducted in a steam or air oven at a few degrees below 100 centimeters. Neither style of jar can be heated directly on the water or steam *bath* with safety to the contents. This needs further trial.

Another slight deviation from the method of last year, which I find convenient, is to retain the cake of insoluble acids in the Erlenmeyer for weighing, using alcohol merely to dissolve into the flask whatever fat acids are on the dried filter. This small amount of alcohol is quickly removed, first on the water bath and then in the oven, with a current of dry air or hydrogen.

By these two changes in the method the entire determination of insoluble acids is effected without any transfer of material from first to last. At the start the filtered, still liquid, and well-shaken fat is weighed into the tared Erlenmeyer; at the end, after the insoluble acids are weighed, they are removed and the flask is dried and tared again to insure against error from "nicking" or corrosion during the operation. However, in the few trials thus far made, the first and last weights of the flask have been practically identical.

The report of the committee was then adopted.

Dr. W. J. Gascoyne then read the report of the Committee on Phosphoric Acid, the president stating that, at the request of the Executive Committee, Dr. Gascoyne had retained the position of chairman on his resignation as State chemist of Virginia.

REPORT OF THE COMMITTEE ON PHOSPHORIC ACID.

As required by the constitution of the association, the Committee on Phosphoric Acid herewith presents its report. The report presents: 1, brief notices of new analytical methods for determining phosphoric acid proposed during the year; 2, results of work done by the committee and members of the association; 3, recommendations for the next year.

C. Mohr (Chem. Zeit., 11, 417–418; Abs. J. Chem. Soc., 1887, p. 884) proposes the following volumetric method for phosphoric acid: Two grams of the phosphatic substance is treated with successive small quantities of 2 per cent. sulphuric acid in a mortar. The residue is added to the extracts, and the whole made up to 100 cubic centimeters and allowed to digest for one hour; 10 cubic centimeters of the filtered solution, corresponding to 0.2 grams, are then treated with potassium ferrocyanide so long as iron is precipitated. After the addition of sodium acetate, the phosphoric acid is titrated in the usual way with a standard solution of uranium acetate. It is stated that the end reaction is not influenced either by an excess of potassium ferrocyanide or by the presence of Prussian blue.

G. Kennepohl (Chem. Zeit., 11, 1089–1091; Abs. J. Chem. Soc., 1888, 321; J. Soc. Chem. Ind., 6, 680) confirms the opinion expressed by Klein (Chem. Zeit., 10, 721; J. Chem. Soc., 1886, 835) as to the practical non-occurrence of iron phosphide in normal Thomas slag. The phosphoric acid may be accurately determined in the following manner: Ten grams of the finely powdered slag are introduced into a 500 cubic centimeter flask, moistened with alcohol to prevent adhesion, and heated in a water bath for at least half an hour with 40 cubic centimeters of HCl. (1.12 sp. gr.) and 40 cubic centimeters of water. After cooling the flask is filled to the graduation mark and the solution filtered. An aliquot part is mixed with ammonium nitrate and molybdic solution without previous removal of the silica. The solution, after heating at about 80° for fifteen minutes, is filtered, and the precipitate washed with water containing 3 per cent. of nitric acid, redissolved in $2\frac{1}{2}$ per cent. ammonia, and then precipitated with magnesia mixture. The presence of silica does not interfere, owing to the ready solubility of ammonium silico-molybdate in the washing water.

If the presence of ferrous salts should tend to cause the separation of molybdic oxide, which dissolves but slowly in ammonia, an addition of nitric acid or bromine, before adding the molybdic solution, will insure the production of a precipitate instantaneously soluble in the ammonia.

Isbert and Stutzer (Zeit. Anal. Chem., 26, 583–587; Abs. J. Chem. Soc., 1888, 191; Chem. News, 57, 211; J. Soc. Chem. Ind., 7, 44; J. Anal. Chem., 2, 193). The method based on the determination of the ammonia in the phospho-molybdate precipitate (Chem. Zeit., 11, 223) is confirmed. A further simplification consists in washing the yellow precipitate with cold water, instead of with ammonium nitrate solution. The ammonium silico-molybdate is soluble in pure cold water, although insoluble in ammonium nitrate solution. On the other hand, the phospho-molybdate requires 10,000 parts of cold water for its solution.

The analysis is conducted as follows:

Five grams of the phosphate are dissolved in hydrochloric acid or aqua-regia, diluted to 500 cubic centimeters and filtered. Fifty centimeters of the filtrate are mixed with an excess of ammonia, acidulated with nitric acid, and the phosphoric acid precipitated with ammonium molybdate. The precipitate is allowed to settle at 60° to 70° C. for fifteen minutes, and then filtered, the supernatent liquid being first passed through the filter, the precipitate repeatedly washed by decantation, then placed upon the filter and washed with water until the wash water amounts to about 250 cubic centimeters. The precipitate is transferred, filter and all, to a $\frac{3}{4}$-liter Erlenmeyer flask, sodium hydroxide added in excess, and the ammonia distilled off into a solution of standard acid, which is then titrated back with baryta water, using rosalic acid as the indicator.

One part of nitrogen in the precipitate corresponds with 1.654 parts of phosphoric acid.

Test analyses with known quantities of phosphate, with and without silicic acid, and comparative tests of phosphatic manures by the above and gravimetric process, show that for commercial purposes the method is sufficiently accurate.

Ch. Malot (Monit. Scient., 1887, 487; J. Pharm., 16, 157–159; J. Chem. Soc., 1887, 1063; J. Soc. Chem. Ind., 6, 563) describes a new volumetric method for P_2O_5 by uranium nitrate. Oxide of uranium gives a green lake with cochineal, which property is utilized for the determination of phosphoric acid.

The phosphate is treated according to Joulie's method, i. e., dissolved in HCl, the phosphoric acid precipitated with citro-magnesium mixture and the precipitate dissolved in dilute nitric acid. A few drops of tincture of cochineal are added, then ammonia until the violet coloration just appears, and this in its turn is made to disappear with one to two drops of nitric acid. The solution is now heated to 100° C.,

5 cubic centimeters of sodium acetate solution added and mixture titrated with uranium nitrate. Each drop of the latter causes a greenish-blue zone, which, on agitation, disappears again. As soon as precipitation is complete, the solution assumes a lasting greenish-blue color, remaining unchanged by excess of the uranium solution. The end reaction is most distinct. By employing a very dilute solution of uranium nitrate, the determination is rendered very exact. The author uses solutions, 1 cubic centimeter of which represents 0.002 gram of phosphoric acid.

A. Emmerling (Zeits. Anal. Chem., 26, 244, Abs. Chem. News, 57, 15; J. Anal. Chem., 2, 228) describes a new method for the determination of soluble phosphoric acid in superphosphates.

The solution of superphosphates, mixed with calcium chloride, is allowed to flow into a standardized solution of caustic soda, to which some phenolphtalein has been added. The following solutions are required :

(1) Sodium hydrate, of which 1 cubic centimeter represents about .005 $P_2 O_5$.

(2) Calcium chloride solution made by dissolving 200 grams $CaCl_2$ in one liter of water.

(3) Phenolphtalein, 1 gram in 500 cubic centimeters of alcohol.

(4) Methyl orange in aqueous solution.

The execution of an analysis is carried on as follows:

Two hundred cubic centimeters of the solution of the sample prepared in the ordinary manner are well mixed with 50 cubic centimeters solution of calcium chloride. One burette is filled with this mixed solution and a second with the soda solution. Of this latter 20, 10, or 5 cubic centimeters (according to the strength of the superphosphate) are measured into a beaker ; 2 cubic centimeters of the phenolphtalein solution are added, and some water, and the superphosphate solution is then run rather rapidly. As soon as the color begins to grow faint the superphosphate liquor is added more gradually, and at last drop by drop, until the redness has entirely disappeared.

The author next measures off again the same number of cubic centimeters of the mixed solution of superphosphate and calcium chloride as used above, dilutes with a little water and mixes with 4 to 6 drops of the methyl orange solution. The liquid is titrated cautiously with the soda solution, finally drop by drop until the reddish color changes to a yellow or orange yellow.

The smaller number of cubic centimeters of soda solution, as consumed in the titration with methyl-orange, is deducted from the number obtained on titrating with phenolphtalein. If we have dissolved 20 grams superphosphate in 1,000 cubic centimeters of water, and make up 200 to 250 cubic centimeters by the addition of 50 cubic centimeters of the calcium chloride solution, and if a is the measured quantity of soda, b the solution of superphosphate and calcium chloride consumed, c the soda used for neutralizing the free acid, and t the standard of the soda expressed as $P_2 O_5$, the percentage of soluble phosphoric acid in the sample is equal to

$$a - c \times t \times \frac{250 \times 1000 \times 100}{200 \times 6.20}$$

This method is applicable to ferruginous superphosphates.

J. Ruffle (J. Soc. Chem. Ind., 6, 327–333; Abs. J. Chem., Soc., 1888, p. 87 ; J. Anal Chem., 1, p. 455) has contributed an interesting paper on moisture and free acids in superphosphate. The loss of moisture, dried to constant weight, at various temperatures varied in one sample from 12.92 per cent. at 38° to 50° C., to 17.93 per cent. at 150°C. Drying at the same temperature, 100° C., for different periods, showed a variation of from 9.97 per cent. for thirty minutes, to 17.15 per cent. for seven hours. The fertilizer dried in its natural state, at 100° C., showed a much greater loss of moisture than when previously rubbed to a fine paste in a mortar.

It is also shown that the soluble phosphoric acid existing in superphosphates is not entirely present as mono-calcium phosphate, and that the exposure of 100° C. drives off more than the true moisture, that is, the adhering uncombined water. It is recommended to determine the moisture in the following manner: Weigh out 2 to 5 grams of the phosphate in its natural state on a double watch glass, place under an air pump over dry calcium chloride, exhaust, then leave for eighteen to twenty-four hours and weigh.

John Clark (J. Soc. Chem. Ind., 7, 311–312) describes a modification of Perrot's (Compt. Rend, 93, 495) for the estimation of phosphoric acid as phosphate of silver.

The objection to Perrot's process, when applied to manures and natural phosphates, are—

(1) The chlorides which may be present will be estimated as phosphates.

(2) The method of separating the phosphates of iron and alumina from the phosphates of lime and magnesia is inaccurate, as it is practically impossible to dissolve out the whole of the prosphate of lime in the manner indicated.

The modifications which the author recommends are—

(1) The solution of the phosphate of silver precipitate in nitric acid, and the titration of the silver with sulpho-cyanide.

(2) The neutralization of the acid solution with caustic soda instead of ammonia, to avoid the presence of an excessive quantity of ammoniacal salt, which affects the results.

(3) The previous precipitation of the iron and alumina as phosphate with acetate of soda containing free acetic acid.

The following is an outline of the process: The phosphate is dissolved either in water, nitric acid, or sulphuric acid, the greater portion of the free acid neutralized with caustic soda, and to the cold solution acetate of soda containing free acetic acid is added in excess. If the addition of the acetate of soda produces a precipitate, this must be filtered off, re-dissolved, and re-precipitated with acetate of soda, as before. The filtrate and washings are added to the previous filtrate, then excess of nitrate of silver, which will give an immediate precipitate of phosphate of silver, $Ag_3 PO_4$, and the free acetic acid is neutralized with caustic soda till there is only a faint acid reaction to litmus paper.

The slightest excess of caustic soda will cause a brown precipitate of oxide of silver, but this oxide of silver dissolves easily on the addition of a few drops of dilute acetic acid.

The precipitate of phosphate of silver, which will contain any chloride which may have been present, is thrown on a filter, thoroughly washed with water, then dissolved off the filter with hot dilute nitric acid, mixed with a little ferric sulphate, and the silver titrated with sulpho-cyanide, as described by Volhard, and calculated to phosphoric acid.

The precipitate of phosphate of iron and alumina is dried, ignited, and weighed, then dissolved in acid, the iron determined volumetrically, calculated to phosphate of iron, and the balance assumed to be phosphate of alumina ; or the oxide of iron may be separated with caustic soda and the alumina in the filtrate weighed as phosphate of alumina, using the precautions recommended by Thomson (J. Soc. Chem. Ind., vol. 5).

Comparative tests on a variety of materials show a close agreement with the molybdic method.

Dircks and Werenskiold (Landw. Versuchs. Stat., 1887,425–453; Abs. J. Chem. Soc., 1888, 628) have tested the various processes employed for the estimation and separation of tri-calcium from mono- and di-calcium phosphate, namely, the various modifications of the ammonium citrate method. They find that although none of the methods give a really satisfactory and exact result, Petermann's process is perhaps the most trustworthy.

C. Schindler (Zeits Anal. Chem., 27, 142–146 ; J. Chem, Soc., 1888, 753, J. Soc. Chem. Ind., 7, 455) describes a volumetric method for the determination of phosphoric acid based upon the formation of a compound of phospho-molybdate of ammonia of definite . composition.

The following solutions are requisite:

(1.) *Molybdic solution.*—To a liter of molyldic acid solution prepared in the usual manner, 30 cubic centimeters of a solution of citric acid containing 500 grams to the liter are added.

(2) Concentrated ammonium nitrate solution containing 750 grams to the liter.

(3) Dilute ammonium nitrate solution containing 100 grams to the liter and 10 cubic centimeters nitric acid.

(4) Magnesia mixture prepared in the usual manner.

(5) Lead solution, 1 cubic centimeter of which corresponds to .04 grams P_2O_5. This is obtained by dissolving 55 grams lead acetate in one liter of water and a little acetic acid.

(6) Molybdate of ammonium solution, 1 cubic centimeter of which corresponds to 1 cubic centimeter lead solution; 25 grams ammonium molybdate are dissolved in one liter and standardized against the lead solution.

(7) Tannin solution, 1 gram tannin is dissolved in 20 to 30 cubic centimeters of water.

The analysis is conducted as follows: 50 cubic centimeters of the nitric acid solution of the phosphate (0.5 gram of substance) is mixed with so much of the concentrated solution of ammonium nitrate that after the addition of the molybdic acid solution the mixture shall contain 25 grams ammonium nitrate per 100 cubic centimeters. Then for each 0.1 gram of P_2O_5, 100 cubic centimeters of the molybdic acid solution is added, and the mixture heated in a water bath to about 58° C. The precipitate is allowed to settle for ten minutes, and the liquid decanted and passed through a filter. The precipitate is washed with 50 cubic centimeters dilute ammonium nitrate solution by decantation, and dissolved in 3 per cent. ammonia solution. The solution, together with that dissolved off the filter, is brought into a quarter-liter flask, 10 to 20 cubic centimeters magnesia mixture added; after shaking, it is made up to 250 cubic centimeters and filtered. Fifty cubic centimeters of the filtrate are taken, acidified with acetic acid, and the liquid made up with boiling water to 300 cubic centimeters. Lead solution is now added from a burette until a small excess of lead goes into solution, and is titrated back with ammonium molybdate solution, using the tannin solution as an indicator. A drop of tannin solution gives a red coloration with ammonium molybdate, which is visible in a solution of 1 in 400,000, whereas lead molybdate gives no coloration; and lead acetate only a greenish coloration.

Comparative determinations on a variety of materials show a close agreement with the magnesia method.

At a meeting of the directors of the Belgian experiment stations, the following method for the estimation of the phosphoric acid soluble in water and in citrate of ammonia in manures was adopted:

Two grams of ordinary superphosphate (or one gram of a rich sample, say over 20 per cent.) are triturated in a mortar with water and thrown upon a filter. The filter is washed with water until about 200 cubic centimeters have passed through; to the filtrate is added 1 cubic centimeter of nitric acid, and the volume of the solution is made up to 250 cubic centimeters. The filter, and the insoluble matter contained in it, is now introduced into a 250 cubic centimeters flask with 50 cubic centimeters of Petermann's alkaline citrate of ammonia solution, and allowed to digest for one hour on the water bath at a temperature of 38° to 48° C. After rapidly cooling the flask is filled up to the mark with water. Of the above filtered liquid (the citrate of ammonia solution), take 50 cubic centimeters and mix it with 50 cubic centimeters of the aqueous solution, acidify with nitric acid and precipitate with molybdic acid solution. The remainder of the operations remain as for the ordinary estimation of phosphoric acid.

Petermann's alkaline citrate of ammonia is prepared by dissolving 500 grams of citric acid crystals in water and mixing with 700 cubic centimeters of ammonia of 0.920 specific gravity. Bring the concentration of the liquid to 1.09 specific gravity at 15° C., and add to each liter 50 cubic centimeters ammonia of .920 specific gravity.

RESULTS OF WORK DONE DURING THE YEAR.

Last October your committee sent to nineteen official and seventeen commercial chemists five samples for phosphoric acid determination, with the request that the determinations be made within the week ending October 29.

The samples were marked as follows:

No. 1. Ground South Carolina phosphate.

No. 2. Ground tankage.

No. 3. Ammoniated superphosphate.

No. 4. Dissolved South Carolina phosphate.

No. 5. Dissolved Navassa phosphate.

The samples were carefully mixed, so as to secure uniformity.

Returns have been received from twenty-seven laboratories, covering the work of thirty-two chemists. The results are given in the following tables:

No. 1 and No. 2.

Analyst.	No. 1—South Carolina phosphate.		No. 2—Tankage.	
	Moisture.	Phosphoric acid.	Moisture.	Phosphoric acid.
E. H. Farrington, Connecticut experiment station		27.67		14.39
T. B. Osborne, Connecticut experiment station		27.85		14.30
A. L. Winton, Connecticut experiment station		27.89		14.33
J. P. Carson, Richmond, Va		28.14		13.95
C. Glaser, Baltimore, Md	.78	28.30	6.78	14.01
W. J. Gascoyne, Baltimore, Md	.93	28.14	6.66	14.20
Lehmann & Mager, Baltimore, Md	.80	28.78	6.03	14.52
C. C. Read, New Bedford, Mass.		27.80		14.91
B. Terne, Philadelphia, Pa		27.68		14.24
Rasin Fertilizer Company, Baltimore, Md	.05	28.20	6.88	14.02
P. B. Wilson, Baltimore, Md	.94	28.22	6.94	14.06
R. H. Gaines, Richmond, Va		28.30		14.71
J. M. Bartlett, experiment station, Orono, Me	.92	28.00	7.14	13.89
S. H. Merrill, experiment station, Orono, Me	.92	27.87	7.14	14.00
W. Frear, State College, Pa	1.37	27.84	7.64	14.40
Clifford Richardson, Washington, D. C		27.67		14.87
Stillwell & Gladding, New York	1.08	28.33	7.00	14.59
C. E. Buck, Wilmington, Del		27.98		14.50
W. Robertson, Charleston, S. C	.83	28.40	6.55	14.71
M. A. Scovell, Kentucky experiment station	1.02	27.87	6.16	14.41
A. M. Peter, Kentucky experiment station	1.00	27.84	6.26	14.15
H. B. Battle, North Carolina experiment station	1.04	28.22	7.08	14.32
A. F. Shireick, Wood's Holl, Mass		28.34		14.03
N. W. Lord, Columbus, Ohio	.97	27.47	7.35	14.06
W. McMurtrie, Champaign, Ill		27.37		14.13
A. E. Knorr, Department of Agriculture, Washington	.76	28.16	6.44	14.23
J. M. McCandless, Atlanta, Ga	1.05	27.78	6.70	14.16
H. Froehling, Richmond, Va		28.08		13.62
P. E. Chazal, Columbia, S. C	.80	28.91	6.70	14.75
Bradley Fertilizer Company (Benson), Boston		27.65		14.91
Bradley Fertilizer Company (Patrick), Boston		28.40		14.35

No. 3.—Ammoniated superphosphate.

Analyst.	Moisture.	Soluble phosphoric acid.	Reverted phosphoric acid.	Available phosphoric acid.	Insoluble phosphoric acid.	Total phosphoric acid.
E. H. Farrington, Connecticut experiment station..	14.59	7.02	1.52	8.54	1.40	9.94
T. B. Osborne, Connecticut experiment station.....	14.82	7.12	.95	8.07	1.84	9.91
A. L. Winton, Connecticut experiment station......		7.11	1.25	8.36	1.36	9.72
J. P. Carson, Richmond, Va.........................	17.21	7.38	1.39	8.77	1.64	10.41
C. Glaser, Baltimore, Md...........................	16.60	6.75	1.41	8.16	1.92	10.08
W. J. Gascoyne, Baltimore, Md.....................	17.15	7.08	1.76	8.84	1.39	10.23
Lehmann & Mager, Baltimore, Md	14.05	7.16	1.35	8.51	1.79	10.30
C. C. Read, New Bedford, Mass.....................	12.00	8.57	1.03	9.60	1.89	11.49
B. Terne, Philadelphia, Pa.........................	14.81	7.16	1.55	8.71	1.72	10.43
Rasin Fertilizer Company, Baltimore, Md..........	16.40	7.00	1.05	8.05	2.20	10.25
P. B. Wilson, Baltimore, Md........................	16.38	7.03	1.02	8.05	2.17	10.22
R. H. Gaines, Richmond, Va........................		7.19	1.14	8.33	2.00	10.33
J. M. Bartlett, experiment station, Orono Me......	12.87	7.10	1.34	8.44	1.49	9.93
S. H. Merrill, experiment station, Orono, Me......	12.87	7.01	1.58	8.59	1.38	9.97
W. Frear, State College, Pa.........................	18.36	6.00	2.27	8.27	1.60	9.87
Clifford Richardson, Washington, D. C	18.10	7.49	.80	8.29	1.59	9.88
Stillwell & Gladding, New York.....................	14.30	7.15	1.22	8.37	1.73	10.10
C. E. Buck, Wilmington, Del........................		7.00	1.36	8.36	1.44	9.80
W. Robertson, Charleston, S. C.....................	15.70	7.29	1.49	8.78	1.58	10.36
M. A. Scovell, Kentucky experiment station......	13.91	6.97	1.17	8.14	1.78	9.92
A. M. Peter, Kentucky experiment station	13.88	7.09	1.27	8.36	1.56	9.92
H. B. Battle, North Carolina experiment station ..	14.39	7.02	.93	7.95	1.78	9.73
A. F. Shirelck. Wood's Holl, Mass.................	16.50	7.00	1.27	8.27	1.52	9.79
N. W. Lord, Columbus, Ohio........................	19.65	7.02	1.42	8.44	1.40	9.84
W. McMurtrie, Champaign, Ill......................		7.07	1.78	8.85	1.45	10.30
A. E. Knorr, Department of Agriculture, Washington	14.06	6.74	1.59	8.33	1.30	9.63
J. M. McCandless, Atlanta, Ga.....................	12.95	7.14	1.20	8.34	1.35	9.69
W. W. Cooke, Burlington, Vt........................		6.99	.93	7.92	1.82	9.74
H. Froehling, Richmond, Va........................		7.19	1.26	8.45	1.52	9.97
P. E. Chazal, Columbia, S. C.......................	15.95	7.48	1.26	8.74	1.56	10.30
Bradley Fertilizer Company (Benson), Boston......		7.20	.99	8.19	1.81	10.00
Bradley Fertilizer Company (Patrick), Boston.....		7.04	1.22	8.26	2.00	10.26

No. 4.—Dissolved South Carolina phosphate.

Analyst.	Moisture.	Soluble phosphoric acid.	Reverted phosphoric acid.	Available phosphoric acid.	Insoluble phosphoric acid.	Total phosphoric acid.
E. H. Farrington, Connecticut experiment station..	9.57	10.95	3.18	14.13	1.47	15.60
T. B. Osborne, Connecticut experiment station.....	9.57	11.04	2.88	13.92	1.77	15.69
A. L. Winton, Connecticut experiment station.....		11.03	3.26	14.29	1.32	15.61
J. P. Carson, Richmond Va.........................	9.77	8.67	5.21	13.88	2.20	16.08
C. Glaser, Baltimore, Md...........................	9.69	11.02	3.13	14.15	1.82	16.17
W. J. Gascoyne, Baltimore, Md.....................	9.41	11.03	3.23	14.26	1.73	15.99
Lehmann & Mager, Baltimore, Md..................	9.28	11.00	3.07	14.07	2.11	16.18
C. C. Read, New Bedford, Mass.....................	9.03	11.30	2.84	14.14	1.76	15.90
B. Terne, Philadelphia, Pa	9.76	10.81	2.68	13.49	2.59	16.08
Rasin Fertilizer Company, Baltimore, Md.........	10.00	11.02	3.20	14.22	1.96	16.18
P. B. Wilson, Baltimore, Md	10.07	11.00	3.24	14.24	1.90	16.14
R. H. Gaines, Richmond, Va........................		11.51	2.24	13.75	2.24	15.99
J. M. Bartlett, experiment station, Orono, Me......	9.12	10.93	2.45	13.38	2.08	15.46
S. H. Merrill, experiment station, Orono, Me......	9.12	10.76	3.18	13.94	1.70	15.64
W. Frear, State College, Pa.........................	10.42	10.84	3.26	14.10	1.70	15.80
Clifford Richardson, Washington, D. C	11.14	11.23	2.28	13.51	2.13	15.64
Stillwell & Gladding, New York.....................	9.53	10.82	3.03	13.85	2.15	16.00
C. E. Buck, Wilmington, Del........................		10.84	3.41	14.25	1.34	15.59
W. Robertson, Charleston, S. C.....................	9.40	11.26	2.55	13.81	2.12	15.93
M. A. Scovel, Kentucky experiment station........	9.65	10.73	2.73	13.46	1.97	15.43
A. M. Peter, Kentucky experiment station	9.55	10.82	2.82	13.64	1.75	15.39
H. B. Battle, North Carolina experiment station ...	9.20	10.94	2.24	13.18	2.19	15.37
A. F. Shireick, Wood's Holl, Mass.................	9.58	10.87	3.31	14.18	1.40	15.58
N. W. Lord, Columbus, Ohio........................	9.53	10.73	3.01	13.74	1.72	15.46
W. McMurtrie, Champaign, Ill......................		11.16	1.21	12.37	2.81	15.18
A. E. Knorr, Department of Agriculture, Washington	9.33	10.63	3.31	13.94	1.28	15.22
J. M. McCandless, At anta, Ga.....................	9.60	10.50	3.12	13.62	2.28	15.90
W. W. Cooke, Burlington, Vt		10.70	2.69	13.39	2.24	15.63
H. Froehling, Richmond, Va........................		11.18	2.75	13.93	1.68	15.61
P. E. Chazal, Columbia, S. C	9.40	11.32	2.55	13.87	1.99	15.86
Bradley Fertilizer Company (Benson), Boston ...		10.80	2.30	13.10	2.31	15.41
Bradley Fertilizer Company (Patrick), Boston.....		11.00	2.54	13.54	2.66	16.20

33

No. 5.—Dissolved Navassa phosphate.

Analyst.	Moisture.	Soluble phosphoric acid.	Reverted phosphoric acid.	Available phosphoric acid.	Insoluble phosphoric acid.	Total phosphoric acid.
E. H. Farrington, Connecticut experiment station...	9. 29	8. 00	7. 38	15. 38	3. 18	18. 56
T. B. Osborne, Connecticut experiment station	9. 29	8. 00	7. 21	15. 21	3. 22	18. 43
A. L. Winton, Connecticut experiment station	......	7. 93	7. 52	15. 35	3. 02	18. 37
J. P. Carson, Richmond, Va	9. 58	6. 78	8. 78	15. 56	3. 39	18. 95
C. Glaser, Baltimore, Md	10. 35	6. 36	8. 82	15. 18	3. 41	18. 59
W. J. Gascoyne, Baltimore, Md...................	9. 15	7. 38	7. 78	15. 16	3. 33	18. 49
Lehmann & Mager, Baltimore, Md	8. 72	6. 78	8. 89	15. 67	3. 52	19. 19
C. C. Read, New Bedford, Mass..................	8. 09	7. 90	7. 27	15. 17	3. 20	18. 37
B. Terne, Philadelphia, Pa	9. 63	6. 91	8. 44	15. 35	3. 77	19. 12
Rasin Fertilizer Company, Baltimore, Md..........	10. 40	6. 05	8. 62	14. 67	3. 86	18. 53
P. B. Wilson, Baltimore, Md....................	10. 35	6. 00	8. 62	14. 62	3. 81	18. 43
E. H Gaines, Richmond, Va	......	7. 52	7. 67	15. 19	3. 52	18. 71
J. M. Bartlett, experiment station, Orono, Mo......	8. 65	8. 05	6. 93	14. 98	3. 61	18. 59
S. H. Merrill, experiment station, Orono, Me	8. 65	7. 08	7. 96	15. 04	3. 42	18. 46
W. Fruar, State College, Pa	11. 91	6. 92	7. 82	14. 74	3. 70	18. 44
Clifford Richardson, Washington, D. C............	15. 80	8. 03	6. 66	14. 69	3. 39	18. 08
Stillwell & Gladding, New York.................	8. 92	6. 93	8. 52	15. 45	3. 21	18. 66
C. E. Buck, Wilmington, Del	......	6. 87	8. 16	15. 03	3. 26	18. 29
W. Robertson, Charleston, S. C.................	8. 55	7. 55	8. 01	15. 56	3. 31	18. 87
M. A. Scovell, Kentucky experiment station	9. 09	8. 06	6. 72	14. 78	3. 52	18. 30
A. M. Peter, Kentucky experiment station.........	9. 27	8. 08	6. 52	14. 60	3. 65	18. 25
H. B. Battle, North Carolina experiment station....	9. 75	7. 46	7. 38	14. 84	3. 34	18. 18
A. F. Shireick, Wood's Holl, Mass	9. 10	7. 33	7. 89	15. 22	2. 93	18. 15
N. W. Lord, Columbus, Ohio	10. 83	7. 47	7. 61	15. 08	3. 13	18. 21
W. McMurtrie, Champaign, Ill...................	......	10. 04	4. 48	14. 52	3. 58	18. 10
A. E. Knorr, Department of Agriculture, Washington	8. 25	6. 78	7. 94	14. 72	3. 34	18. 06
J. M. McCandless, Atlanta, Ga	9. 50	9. 88	5. 65	15. 53	2. 95	18. 48
W. W. Cooke, Burlington, Vt	......	7. 34	7. 82	15. 16	3. 68	18. 84
H. Froehling, Richmond, Va	......	7. 39	7. 67	15. 06	3. 30	18. 36
P. E. Chazal, Columbia, S. C...................	8. 95	8. 25	7. 65	15. 90	3. 10	19. 00
Bradley Fertilizer Company (Benson), Boston......	......	7. 11	7. 05	15. 06	3. 54	18. 60
Bradley Fertilizer Company (Patrick), Boston......	......	7. 95	7. 05	15. 00	3. 70	18. 70

Averages.

	Number of determinations.	Average.	Highest.	Lowest.	Difference.
Sample No. 1:					
Moisture	17	. 95	1. 37	. 76	. 61
Phosphoric acid..............	31	28. 07	28. 78	27. 47	1. 31
Sample No 2:					
Moisture	17	6. 79	7. 64	6. 03	1. 61
Phosphoric acid..............	31	14. 31	14. 01	13. 62	1. 29
Sample No. 3:					
Moisture	24	15. 40	19. 65	12. 60	7. 05
Soluble phosphoric acid......	32	7. 11	8. 57	6. 00	2. 57
Reverted phosphoric acid	32	1. 31	2. 27	. 80	1. 47
Available phosphoric acid	32	8. 42	9. 60	7. 92	1. 68
Insoluble phosphoric acid	32	1. 65	2. 20	1. 30	. 90
Total phosphoric acid........	32	10. 07	11. 49	9. 63	1. 86
Sample No. 4:					
Moisture	24	9. 61	11. 14	9. 03	2. 11
Soluble phosphoric acid......	32	10. 88	11. 51	8. 67	2. 84
Reverted phosphoric acid	32	2. 93	5. 21	1. 21	4. 00
Available phosphoric acid	32	13. 81	14. 29	12. 37	1. 92
Insoluble phosphoric acid....	32	1. 95	2. 81	1. 28	1. 53
Total phosphoric acid........	32	15. 76	16. 20	15. 18	1. 02
Sample No. 5:					
Moisture	24	9 66	15. 80	8. 09	7. 71
Soluble phosphoric acid......	32	7. 52	10. 04	6. 00	4. 04
Reverted phosphoric acid	32	7. 59	9 05	4. 48	4. 57
Available phosphoric acid	32	15. 11	15. 90	14. 52	1. 38
Insoluble phosphoric acid	32	3. 40	3. 86	2. 93	. 93
Total phosphoric acid........	32	18. 51	19. 19	18. 06	1. 13

As further illustrating the differences in the results the following table is presented:

	Percentage of results differing from the average not more than—								
	.10	.25	.50	.75	1.00	1.25	1.50	1.75	2.00 and over.
Sample No. 1:									
Moisture	19	61	91	100					
Phosphoric acid	29	52	90	100					
Sample No. 2:									
Moisture	59	94	100						
Phosphoric acid	23	47	70	85	100				
Sample No. 3:									
Moisture			4	17	29	46	58	66	100
Soluble phosphoric acid	50	78	94	94	94	94	97	100	
Reverted phosphoric acid	41	63	97	97	100				
Available phosphoric acid	38	63	97	97	100				
Insoluble phosphoric acid	22	69	94	100					
Total	28	59	97	97	97	97	97	100	
Sample No. 4:									
Moisture	37	54	87	92	96	96	96	100	
Soluble phosphoric acid	31	72	94	97	97	97	97	97	100
Reverted phosphoric acid	16	44	84	94	94	94	94	97	100
Available phosphoric acid	22	37	91	97	97	97	100		
Insoluble phosphoric acid	13	50	78	97	100				
Total	9	53	96	100					
Sample No. 5:									
Moisture	12	17	20	63	67	83	87	91	100
Soluble phosphoric acid	12	28	47	84	84	87	90	93	100
Reverted phosphoric acid	16	34	53	63	75	88	94	94	100
Available phosphoric acid	34	50	88	97	100				
Insoluble phosphoric acid	28	66	100						
Total	34	59	94	100					

The committee recommends the continuance of the present general method for the determination of phosphoric acid, subject to such modifications as the association may determine.

Respectfully submitted,

W. J. GASCOYNE,
Chairman.

The report being open for discussion, the president called attention to the dependence of amount of soluble phosphoric acid on the size of filters used during its extraction, and thought it would be well for the committee to recommend a definite size of 9 centimeters.

Professor Frear spoke of differences due to variations in moisture in different laboratories.

Mr. Gascoyne said the variations in moisture was not proportional to the variations in phosphoric acid, and that such differences as 12 and 19 per cent. could not be due to anything but carelessness, the highest determinations of moisture being accompanied by highest percentages of phosphoric acid.

Professor Lupton, on examining the results, said he could see no way to harmonize variations.

The president thought the publication of these results would be most beneficial to the association, and that results another year would be sent in with greater care.

The president then introduced the subject of the use of pumps for extracting soluble phosphoric acid, considering it an advantage where much work was to be done.

Mr. Gascoyne spoke of quick work with molybdate and magnesia solutions, allowing them to stand but five to ten minutes for the yellow precipitate and not more than fifteen for magnesia precipitate.

Mr. Gaines's and Mr. Richardson's experiences confirmed his observations.

On motion of Dr. Wiley, further consideration was postponed till afternoon, and then by unanimous consent he offered certain amendments to the constitution, reported later to the convention, and asked for the appointment of a committee of three under the constitution to consider them.

The president appointed Dr. Wiley, Professor Stubbs, and Professor Voorhees.

On Professor Meyer's motion the convention then adjourned till 2 p. m.

AFTERNOON SESSION, THURSDAY.

The convention was called to order by the president at 2 p. m.

The discussion of the report of the committee on phosphoric acid being in order, it was directed that the method for the coming year provide for the use of Schleicher and Schüll No. 589 filters, 9 centimeters in diameter, in the determination of soluble phosphoric acid, and that the preliminary washing be by decantation from a beaker, using small quantities of water, as described by Messrs. Gascoyne and Voorhees, who considered the use of the pestle only necessary with the very high grade phosphates in New England.

After some argument the use of a pump in determining soluble phosphoric acid was left to the judgment of the analyst.

Dr. Gascoyne then called attention to the method of Isbert & Stutzer for the estimation of phosphoric acid, and gave the following determinations, comparing it with the official method. The method was carefully followed, except that half normal sulphuric acid and caustic soda were used, and cochineal as the indicator.

	Association method.	Isbert & Stutzer's method.
Acid phosphate	16.10	16.03
South Carolina phosphate	26.43	26.58
Tankage	11.16	11.19
Steamed bone	30.15	30.07
Ammoniated superphosphate	10.72	10.68
Acid phosphate, total	15.77	15.74
Acid phosphate, soluble	9.63	9.57
Acid phosphate, insoluble	2.26	2.20

He also called attention to the time required for the complete precipitation of the molybdic acid and magnesia precipitates.

The following table is the result of some experiments on this subject. The phosphate was dissolved in the usual manner, filtered, ammonia added in slight excess, acidified with nitric acid, heated to about 80° or 85° C., the molybdic acid solution added, and the solution well stirred. In the first set of experiments the molybdic precipitate was allowed to stand for five minutes, filtered, magnesia mixture added, and allowed to stand two hours. In the second set, the molybdic precipitate was filtered in ten minutes. In the third set, the molybdic precipitate was allowed to stand one hour, and the magnesia precipitate for fifteen minutes. In the fourth set the magnesia precipitate was allowed to stand thirty minutes. In the fifth set the molybdic acid precipitate stood five minutes, and the magnesia precipitate fifteen to twenty minutes.

[Acid phosphate containing 15.82 by the official method.]

Molybdic precipitate five minutes, magnesia precipitate two hours........................... 15.78
Molybdic precipitate ten minutes magnesia precipitate two hours.............................. 15.80
Molybdic precipitate one hour, magnesia precipitate fifteen minutes.......................... 15.80
Molybdic precipitate one hour, magnesia precipitate thirty minutes........................... 15.82
Molybdic precipitate five minutes, magnesia precipitate fifteen to twenty minutes............ 15.79

	Official method.	Molybdic precipitate, five minutes, magnesia precipitate, 15 to 20 minutes.
Acid phosphate, total	16.10	16.07
Acid phosphate, soluble	10.22	10.24
Acid phosphate, insoluble	2.12	2.08
Ground bone	22.14	22.17
Tankage	9.93	9.90

The report was then adopted and the committee directed to insert the modifications in the method.

Dr. Wiley then introduced, by permission, Prof. F. W. Clarke, of the U. S. Geological Survey, who addressed the association on the desirability of forming a national chemical society, and suggested the appointment of a committee to represent the association in any steps that might be taken, and this being agreed to, the president appointed Dr. H. W. Wiley, of Washington; Prof. W. C. Stubbs, of Louisiana; and Prof. E. A. von Schweinitz, of North Carolina.

Dr. Wiley then exhibited one of Abbe's large refractometers and read the following paper:

REFRACTIVE INDEX OF BUTTER FAT.

By H. W. WILEY.

Heretofore very little attention has been given to the refractive index of butter fat in studying the properties of that substance. Having had occasion this year to examine the refractive index of various fats and oils used in the adulteration thereof, I extended the examination to twelve samples of butter fat from butter bought in the open market. The object of the investigation was twofold: First, to determine the mean refractive index; and, second, the rate of variation of the refractive index for rise or fall of temperature. It is, of course, obvious that the refractive index of butter can not be read after it solidifies, and therefore the minimum temperature at which the refractive index can be determined is not much below 30°, although the butter fat may be kept some time even at 25° without solidification. A very warm room where the temperature was reasonably constant was selected in the determination of the refractive index at the lower temperature. For the higher temperature the hot room of a Turkish-bath establishment was used. The temperatures at which the reading was made and the observed index of refraction, the rate of variation for each degree for each sample, and the mean rate of variation for each degree of the twelve samples a e given in the following table:

Number.	To.	Index.	Rate for each degree.	Number.	To.	Index.	Rate for each degree.
1..............	32 / 54. 6	1.4535 / 1.4497	.000169	7..............	32. 2 / 56. 6	1.4510 / 1.4495	.000186
2..............	31. 9 / 56. 2	1.4535 / 1.4495	.000165	8..............	32. 4 / 54. 8	1.4510 / 1.4505	.000157
3..............	31. 8 / 55. 0	1.4530 / 1.4485	.000194	9..............	32. 2 / 55. 0	1.4516 / 1.4500	.000158
4..............	30. 8 / 54. 8	1.4530 / 1.4495	.000146	10..............	31. 8 / 56. 2	1.4530 / 1.4490	.000164
5..............	32. 3 / 55. 8	1.4530 / 1.4495	.000153	11..............	31. 6 / 56. 8	1.4540 / 1.4495	.000186
6..............	33. 2 / 55. 4	1.4535 / 1.4888	.000203	12..............	31. 8 / 56. 2	1.4528 / 1.4485	.000171

The reading of the instrument with distilled water at 18° = 1.3305. Having determined the rate of variation for butter fats, in order to reduce the reading to any standard temperature it is simply necessary to use the mean factor and add or subtract the result of the multiplication as the case may be. For instance, take the case of the first sample. The refractive index at 32° is 1.4535; suppose it is wished to reduce this to a temperature of 25°; the difference in temperature between 32° and 25° is 7°, the mean rate of variation for each degree is .000171, for 7° it would be .001197. Since the index of refraction would be higher at 25° than at 30° we add this number to the number obtained above. Then 1.4535 + .001197 = 1.4547 = index of refraction of that sample at 25°. The index of refraction of butter fat is distinctly lower than that of cotton-seed oil, lard oil, olive oil, and linseed oil. The rate of variation for change of temperature is also less for that of butter fat than the substances mentioned.

On the conclusion of Dr. Wiley's paper a practical demonstration of the working of the lactocrite was given by Mr. A. E. Knorr, after which the convention adjourned until Friday at 10 o'clock.

FRIDAY MORNING.

The convention was called to order at 10 o'clock by the president.

Dr. Farrington called attention to the fact that water was as suitable for washing the yellow molybdate precipitate as ammonic nitrate.

Isbert and Stützer had observed this, and Dr. Gascoyne said that his experience confirmed it.

On motion, the committee on phosphoric acid was directed to insert this as an alternative in the methods.

Dr. Crampton, in the absence of the chairman, Professor Rising, presented the report of the committee on fermented liquors, as contained in the following letter:

OFFICE OF STATE ANALYST,
Berkeley, July 18, 1888.

DEAR SIR: I hope you will not think that I have intentionally been wanting in respect to the other members of the committee on "fermented liquors" because I have till now failed to communicate with you. Early in the year we began some original work upon artificial colors, etc., and it was my expectation to have had some results that I could have submitted to you for your approval before the close of the year, but our work was interrupted by other analyses, and at the last moment the results were not sufficiently satisfactory to offer to the public. However, I believe we are on the right track, and that we may yet be able to recognize many of these artificial colors in a very short time. I had until a few weeks ago fully expected to have been present at the August meeting of the association. I had planned to have spent a few weeks in your laboratory before the annual meeting, and then and there we could have made up our report. This was my plan, but at the last moment I find myself unable to carry it out, and now am unable to make any suggestion. If the association would grant us another year I would promise to be on hand, and I think I could also promise to spend some time in your laboratory, when we could perfect our report. However, if you can make a report I would not in the least hamper you, but wish you to feel at liberty to do so. I have written Professor Wiley and Mr. Clifford Richardson my views, which are intended as an explanation to the association for the failure on my part to do my duty in the matter.

We have proved beyond all question, as I think, the presence of boracic acid in many unadulterated California wines. In addition to the test with tumeric paper we have obtained the flame-test so decided that there can be no doubt. I have just received a new spectroscope, and I shall hope by its aid to settle this point beyond all question. Our method of work is as follows: 50 cubic centimeters to 100 cubic centimeters of wine are evaporated in a platinum dish, ignited, and burned to ash. Part of this ash is transferred to a platinum spoon; such an one as is used for blow-pipe work answers very well. A few drops of the strongest H_2SO_4 (I have used a 96–98 per cent. acid), then alcohol is added and then lighted and immediately blown out, and relighted and again extinguished. The first flash will show the acid very distinctly if it is present.

The adulterations that we have had to look for have been easily detected. The wine in all cases was taken direct from the wine-maker. It was taken upon its sale to the wine-dealer, who had required a certificate of purity before accepting it. Naturally, only pure wine or wine that was believed to be pure would be offered. The only kind of adulteration believed to have been practiced by wine-makers is, first, the use of antiseptics, salicylic acid, sulphurous acid, and boracic acid, coloring matters, alkalis (K_2CO_3) to neutralize excess of acid, etc. The watering of wine (stretching) belongs to the dealer's art, etc. We have not had occasion to investigate this branch of the subject. We examine for plaster, bases added, antiseptics, artificial colors, copper, lead, zinc salts, alcohol, free acid (volatile and non-volatile), solid residue, ash (soluble and insoluble), alkalinity of the soluble portions, sugar, glucose, glycerine, etc., tannic acid, cream of tartar, etc. Some samples of eastern wine were clearly "manufactured."

I have been obliged to go into the country on account of my family and am now in the midst of the Napa Valley wine region and am making the acquaintance of the wine men here, and hope to gain their confidence so that they will consult me in regard to their troubles, etc. There is a strong sentiment among the wine-makers against the use of any agent that could be called an adulterant, and I find them very anxious to maintain the reputation they already have for purity. The wine-dealer is the man we have to fear; he is ready to do almost anything and to get ahead of the chemist if he can. The Viticultural Commission has sent a special agent to Paris. They have instructed him to go about and get as many of the practices of wine makers and dealers as possible. I find great difficulty in getting a set of samples of wine-coloring substances. They are only sold in original packages, etc., under various names. I shall hope through our special agent to get a pretty complete set, and then if I can identify them and give them their true scientific names, we will have a surer base to stand upon.

I regret that I can not attend the meeting this year. If the convention will continue the committee I think we can get to work very early the coming year, and present to the next convention a valuable contribution to the subject. May I ask you to present my apology to Professor Fellows. I do not know his address, nor can I get it till I return to Berkeley.

Very truly, yours,

W. B. RISING.

Dr. C. A. CRAMPTON,
Agricultural Department, Washington, D. C.

Dr. Crampton added that he thought it advisable to adopt as provisional methods those which he had collated from foreign sources and published in Part III of Bulletin No. 13 of the Division of Chemistry of the United States Department of Agriculture.

Mr. Richards suggested that the committee examine into the subject of alcohol tables and recommend some standard that might be brought into common use, as the English, French, and American are now all different, and since the recent investigation of Squibb all shown to be more or less incorrect.

The suggestions of the committee were then adopted.

Prof. Myers, on behalf of the committee on potash, then presented their report as follows:

REPORT OF THE COMMITTEE ON POTASH.

The committee on the determination of potash would respectfully submit: That since the last meeting of this association little has appeared in chemical literature relating to the determination of potash, so far as we know, which calls for notice here. The method of Lindo as modified by Gladding appears to give universal satisfaction. The New Jersey Agricultural Experiment Station (Report for 1887, p. 175) has determined the potash in forty-eight mixed fertilizers by this method, with certain slight modifications, and also by the Stohmann method. The greatest difference in the results in any case was .27 per cent.; the average difference .08 per cent. In two-thirds of the determinations the differences were not over .10 per cent., and in only one case was the difference over .15 per cent. Stohmann's method gave on the average slightly higher results than the other. The modified Lindo-Gladding method used at the New Jersey station is as follows:

"The solutions were made as described in the [official method] but previous to the addition of ammonium hydrate 1 cubic centimeter of ammonium oxalate was added; the solution was then filled to the mark, thoroughly mixed and filtered; an aliquot part was transferred to a porcelain dish of approximately 60 cubic centimeters capacity. From this dish the material was not removed until the double salt of potassio-platinic chloride was ready to be brought upon the Gooch filter. After the determination was removed from the bath the precipitate was moistened with water, and then washed with 80 per cent. alcohol till all traces of platinic chloride were removed. It was then washed with the saturated ammonium chloride solution recommended. It was found in this laboratory that better results could be secured by adding this solution to the precipitate in the porcelain dish, and washing by decantation until all salts other than the potassio-platinic chloride were removed. Since a relatively small amount only of the potassio-platinic chloride is necessary to saturate the ammonium chloride solution, these washings were all rejected, the time saved and the greater accuracy secured compensating for the loss of larger amounts of the solution.

After this washing has been completed, the double salt was transferred to the Gooch filter, washed with pure alcohol, dried and weighed. The salt was then dissolved in hot water, and the final weight of the potassio-platinic chloride secured by difference."

SAMPLES FOR ANALYSIS.

Following the custom adopted by the association, very carefully prepared samples were sent out by the committee with the request that they be analyzed and the results reported to the committee. Twenty-one samples were sent, but reports were received from only nine laboratories or analysts, the most of the others informing the committee that, owing to reorganization and changes taking place due to the influence of the "Hatch bill," they could not find the time to complete the work.

The samples sent out by the committee were numbered from one to five inclusive, and consisted of the following compounds: No. 1, chloride of potassium; No. 2, sulphate of potassium; No. 3, an ammoniated superphosphate mixed with No. 2 in such proportions as to give theoretically 13.50 per cent. of $K_2 O$; No. 4 was calcined kainite; No. 5, an ammoniated superphosphate containing theoretically 5.2 per cent. of $K_2 O$.

Nos. 1 and 2 proved not to be C. P., though bought for C. P. salts.

Table giving the results of the determination of potash in committee's sample.

No. 1.	No. 2.	No. 3.	No. 4.	No. 5.
E. H. Jenkins, analyst, Connecticut Experiment Station.				
....	....	....	[1]17.86	Lost.
....	....	....	[1]17.80	....
[3]53.30	53.78	13.51	[2]17.89	....
53.16	53.84	13.51	[2]17.84	....
53.23	53.81	13.51	17.847	....
William Frear, analyst, Pennsylvania Experiment Station.[4]				
[5]53.34	54.11	13.53	19.28	5.33
Dr. Caldwell's laboratory, Cornell University.				
53.60				
53.77				
53.52	53.64	13.15	17.76	5.21
53.70	53.66	13.24	17.78	5.18
53.647	53.65	13.195	17.77	5.195
Dr. Battle, analyst, North Carolina Experiment Station.				
53.44	[6]51.09	14.49	19.79	6.00
53.44	53.71	14.12	20.26	5.82
53.44	53.90	14.305	20.025	5.91

No. 1.	No. 2.	No. 3.	No. 4.	No. 5.
Dr. Lupton, analyst, Alabama Experiment Station.				
54.27	51.53	13.30	19.68	4.90
54.20	51.87	13.47	19.67	4.61
54.235	51.70	13.385	19.675	4.755
Dr. Crampton, analyst, Department of Agriculture, Washington.				
A.O.A.C	A.O.A.C	A.O.A.C	A.O.A.C	A.O.A.C
53.73	53.48	13.70	17.20	4.96
53.81	53.68	13.82	17.20	4.99
53.75	....	13.90	17.26	....
53.60	....			
53.52	....			
53.682	53.58	13.806	17.22	4.975
L. G.	L. G.	L. G.	L. G.	L. G.
54.11	54.10	13.62	17.32	5.13
54.06	51.02	13.72	17.32	5.30
....	51.85	13.78	....	
54.085	53.993	13.706	17.32	5.21
J.A. Myers, analyst, Mississippi A. and M. College.				
53.30	....	13.52	[7]17.30	....
....	....	13.64	17.24	5.16
....	....	13.58	17.27	....

[1] and [2] Different chemists.
[3] Not C. P. Contains Na. Mg. $S O_4$.
[4] Averages of results.
[5] Not C. P.
[6] Different weighings.
[7] Not satisfactory to analyst, but not time to repeat.

Table giving the results of the determination of potash in committee's sample—Continued.

No. 1.	No. 2.	No. 3.	No. 4.	No. 5.
Dr. B. V. Herff, analyst, Mississippi A. and M. College.				
A.O.A.C.	A.O.A.C.	A.O.A.C.	A.O.A.C.	A.O.A.C.
....	....	13.74	17.64	5.36
....	....	13.70	17.74	5.26
....	....	13.72	17.62	5.20
....	....	....	17.92	5.12
....	....	....	17.76	....
....	....	....	17.58	....
....	....	....	17.50	....
....	....	....	17.42	....
....	....	13.72	17.647	5.234
L. G.	L. G.	L. G.	L. G.	L. G.
53.44	53.56	13.50	17.40	5.16
53.36	53.48	13.48	17.40	5.34
53.36	53.32	....	17.36	5.56

No. 1.	No. 2.	No. 3.	No. 4.	No. 5.
Dr. B. V. Herff, analyst, Mississippi A. and M. College—Continued.				
L. G.	L. G.	L. G.	L. G.	L. G.
....	....	....	17.32	5.22
....	....	....	17.28	....
53.386	53.433	13.49	17.352	5.32
B. W. Kilgore, analyst, Mississippi A. and M. College.				
53.70	....	13.87	[1]17.80	5.96
....	....	13.91	17.08	5.83
....	....	13.89	17.41	5.895

	No. 1.	No. 2.	No. 3.	No. 4.	No. 5.
Average...	53.404	53.524	13.605	18.078	5.295
Maximum..	54.27	54.11	14.49	20.26	6.03
Minimum..	53.16	51.53	13.15	17.08	4.61
Difference...	1.11	2.58	1.34	3.18	1.39
Number of analyses..................................	23	17	24	33	22

[1] Not satisfactory to analyst, but not time to repeat.

Table giving moistures of committee's samples of potash.

Name of analyst.	No. 1.	No. 2.	No. 3.	No. 4.	No. 5.
	Per cent.	Per cent.	Per cent.	Per cent.	Per cent.
Dr. Caldwell	0.065	0.00	4.40	1.25	5.80
Dr. Frear, Pennsylvania Experiment Station......	0.38	0.11	6.29	1.67	7.61
Connecticut Experiment Station..................	0.27	0.21	4.19	1.07	(*)
J. A. Myers[†]......................................	0.3	0.15	4.3	1.30	7.50

* No. 5 broken while in shipment.   † Taken at time of bottling samples.

NOTE.—Other analysts failed to report the moisture.

In giving the results of the analyses the usual custom of giving the averages of all of the results is followed, though the committee does not claim that it represents fairly the work of the chemists of the country. It is hoped, however, that the study of the results of these analyses may lead to more careful work by the various analysts in determining this very difficult element. It is true that among the analyses may be found some results which arise from inexperience, but a perusal of the table shows that some of the very best known analysts in the country differ by more than 0.5 per cent. upon the analysis of the same sample, or as nearly the same sample as it is possible to have mixed and shipped. The most serious difficulty appears to exist with those samples containing sulphate of potassium, i. e., Nos. 2 and 4. In the analyses of these compounds we have some really extraordinary results, a variation of more than 2.5 per cent. occurring between the extremes.

In the cases of at least two of these the attention of the analysts was called to the variation, and the work was at once reviewed. It could not be explained upon the theory of variation of moisture in the samples, as the determinations of moisture reported by the analysts came almost within the "working error." In reviewing these, different chemists in the same laboratories were placed upon them, and the results arrived at, in the judgment of the chief analysts, justified the report made at first by the chief. Every precaution having been taken to have the samples the same, we are driven to the conclusion that there is some minor variation in the methods practiced in the respective laboratories, or else the scheme now in use is inadequate to overcome the difficulties of the case.

In the cases of the other samples the variations from the mean are not so striking, but are still too considerable to be satisfactory. An analyst finding the minimum per cent. and one finding the maximum per cent., reporting upon any one of these samples would differ by more than 1 per cent., which in a large business transaction based upon such analyses would be quite a serious matter.

The results in the judgment of your committee are highly unsatisfactory. The sources of error in the determination of potash seem not yet to be fully understood by our analytical chemists, and demand the most serious study. This association has been working upon this problem for several years, and appears to have made little or no progress towards removing the difficulties.

The various analysts in any one laboratory appear to agree reasonably well with one another in most cases, but their results frequently do not bear comparison with those obtained elsewhere. In some cases, perhaps, care has not been observed to apply proper corrections for impure $PtCl_4$ or other sources of error which might be corrected by checking the work of the laboratory by blank determinations with pure KCl.

Your committee would recommend the continuation of the present schemes for determination of potash, with the modification proposed by the New Jersey Experiment Station.

Respectfully submitted.

JOHN A. MYERS, *Chairman.*
WILLIAM FREAR.

The report being open to discussion, Dr. Gascoyne called attention to the recent experiments of Lindo, contained in Chemical News, 56, 103, approving the association's method.

Professor Scovell said he had some difficulties, and Mr. Richardson spoke of the presence at times in the case of superphosphates of sulphate of lime in the final double salt.

Mr. Voorhees then described the modification practiced at the New Jersey Agricultural Experiment Station, as given in the eighth annual report, p. 175, and said that they found the Lindo-Gladding method most desirable when a little oxalate of ammonia was used to remove lime, and the evaporation was not carried too low, and referred to the large series of results which he had published in the report referred to.

If the evaporation were carried too far, solids might form which would be insoluble in alcohol, in which case a small quantity of water should first be added, and then the strong alcohol. The greatest error in inexperienced hands lies in this point.

Mr. Chazal said he had also found error in not evaporating low enough.

Mr. Voorhees called attention to the fact that the presence of ammonia in the air of the laboratory produced no effect with the Lindo method, as the ammonia double salt would afterward be dissolved out by the wash, and then moved the addition to the association method of the use of oxalate with the ammonia and evaporation to dryness with HCl, taking up with water, repeated evaporation, and taking up with alcohol.

This was adopted.

Professor Scovell then presented the report of the committee on nitrogen, as follows :

REPORT OF THE COMMITTEE ON NITROGEN.

Your committee on nitrogen have the honor to present the following report of the work which has been accomplished during the year :

ABSTRACT OF ARTICLES WHICH HAVE APPEARED DURING THE YEAR ON THE DE-
TERMINATION OF NITROGEN IN FERTILIZERS.

[Compiled, by request, by Dr. E. H. Jenkins, Connecticut Agricultural Experiment Station, and A. M. Peter, Kentucky Agricultural Experiment Station.]

The Kjeldahl method.

Lenz (Fres. Zeitschr. Analyt. Chem., 26, p. 590) has tested the effects of omitting oxidation with potassium permanganate after boiling the substance to be analyzed with sulphuric acid. He used a flask of 100 cubic centimeters' capacity, .2-.7 gram of substance, 10 cubic centimeters of oil of vitriol, and .5 gram metallic mercury. The boiling was continued until the substance was clear and colorless. In one series of determinations no permanganate was added; in another it was added as recommended by Kjeldahl. In all cases higher results were obtained when permanganate was used; generally the differences were small, but in some cases amounted to .33 per cent.

Ulsch (Zeitschr. für das gesammte Brauwesen, 1887, p. 3 ; Fres. Zeitschr., 27, 73) calls attention to the fact that if platinum chloride is added to hasten oxidation when the substance is boiled with oil of vitriol, as recommended by him in a previous paper, an excess must be avoided. Too much platinum chloride retards instead of hastening oxidation, and if the boiling is continued too long, platinum may destroy ammonia by "catalytic action," and so cause loss of nitrogen. He also recommends the use of iron sulphate to destroy mercuro-ammonium compounds formed when mercury is used to hasten oxidation. It has the advantage over potassium sulphide that it can be added to the acid before neutralizing it, thus avoiding danger of losing nitrogen.

Dafert (Landwirtschaft. Versuchs-St., 34, 311: also Fres. Zeitschr. für Analyt. Chem., 27, 222) has studied the chemical reactions involved in the method, and the question how generally applicable the original method of Kjeldahl is, and whether it can be so modified as to make it available for determining nitrogen in all classes of compounds. His conclusions are briefly as follows :

(1) Sulphuric acid withdraws from the nitrogenous organic matter the elements of water and of ammonia, and from them forms ammonia.

(2) The sulphurous acid which is evolved regularly reduces the nitrogenous compounds; but this effect is insignificant compared with that mentioned above.

(3) The addition of organic compounds (sugar, etc.) to nitrogenous matters delays the formation of ammonia, except in cases where it changes a nitrogenous substance, which is volatile or easily decomposed, into one which is less easily attacked by sulphuric acid.

(4) Potassium permanganate, when brought into the hot mixture, destroys the organic compounds still remaining. A part or all of the nitrogen in such compounds forms ammonia. In quantitative work, when the previous digestion with acid has been carried far enough, and the permanganate is added carefully, all the nitrogen is thus converted into ammonia.

(5) His study of the effect of adding metals or metallic oxides leads to the conclusion that the addition of a metal to a substance which is easily oxidized or is easily decomposed by sulphuric acid may cause loss of nitrogen by a too rapid oxidation during the formation of ammonia, and in this way the process is shortened by sacrificing accuracy. Only in special cases, where the substance is very stable, is it safe to use platinum chloride to hasten oxidation, as recommended by Ulsch.

(6) He finds that nitrogen can be accurately determined by the original Kjeldahl method in all amides and ammonium bases, pyridin and chinolin bodies, alkaloids, bitter principles, albuminoids, and allied substances; also most likely in the indol derivatives.

In the following substances nitrogen can not be determined satisfactorily by Kjeldahl's original method.

All nitro-, nitroso-, azo-, diazo-, hydrazo-, and amidoazo- bodies, compounds of nitric and nitrous acids, the hydrazines, and probably cyan- compounds.

For the determination of nitrogen in these classes of compounds probably no general rule can be given, but the peculiarities of each group must be studied as Jodlbauer has studied the special modifications required by nitrates.

E. Waller and H. C. Bowen (Jour. Analy. Chem., 11, 293) give a sketch of the history of the method, and mention modifications which they have adopted, containing nothing essentially new, except that they evaporate off most of the sulphuric acid used in the oxidation of the substance and heat some time after adding the permanganate; an operation which seems hazardous. To the paper is appended a list of the papers which have been published on the Kjeldahl method.

The New Jersey Agricultural Experiment Station (Report N. J. Expt. Sta., 1887, p. 169) has compared, in the case of nineteen fertilizers, the results obtained by the modification of Kjeldahl's method as described by Scovell with those obtained by the absolute method as described in the Report of the Connecticut Experiment Station for 1879. The results showed very satisfactory agreement. The largest difference was .14 per cent.; the average difference, .06 per cent.; in eight cases the absolute method gave on the average .05 per cent. more nitrogen; in eleven cases the modified Kjeldahl method gave on the average .06 per cent. more nitrogen.

E. H. Farrington (Report Conn. Expt. Sta., 1887, p. 126) reports comparative determinations of nitrogen in thirty-four samples of mixed fertilizers containing nitrates by the Jodlbauer-Kjeldahl method and the absolute method as described in the reports of the same station for 1878 and 1879. His conclusions are as follows:

"An inspection of these results shows that in 63 per cent. of the cases the difference between the two methods was not over 0.1 per cent., and in 83 per cent. of the cases not over .15 per cent. The greatest difference was .21 per cent."

"The plus differences are 14 in number; the minus differences, 20. The average of the former is .066; of the latter, .085. It is evident, then, that the two methods are about equally accurate."

Otto Shönherr (Chem. Zeit., 12, 217) applies the azotometer to the Kjeldahl determination. The digestion flasks are graduated by a mark on the neck to a convenient volume (150 cubic centimeters) and after the oxidation has been effected in the usual way the contents of the flask are diluted, neutralized approximately, made up to the mark, and mixed, and an aliquot part (50 cubic centimeters) decomposed in the azotometer with 50 cubic centimeters hypobromite solution. Results accurate.

R. Meldola and E. R. Moritz (Jour. Soc. Chim. Ind., 7, 63) purify sulphuric acid from ammonium sulphate by adding about .05 gram potassium nitrate per 10 cubic centimeters to the acid and heating about two hours.

Other methods.

Houzeau (Pharm. Centr. 28, 627, 628, also Jour. Analyt. Chem. 2, 354) proposes a mixture to be used in determining nitrogen, for the conversion of nitrogen in any combined form into ammonia. Equal weights of sodium acetate and thisulphate are melted in their water of crystallization and allowed to solidify. The mixture is powdered and kept in well-stopped bottles. The combustion is made as follows: In the posterior end of the combustion tube are placed 2 grams of the above mixture and 2 grams of coarsely-pulverized soda-lime, and then a layer of soda-lime a few centimeters long.

The finely-powdered substance to be analyzed (.5 gram if rich in nitrogen, 10–25 if a soil) is well mixed with 10 grams of the salt mixture and 10 grams of soda-lime, and filled into the tube, followed by soda-lime and an asbestus or glass-wool plug. The combustion is made as usual, and the mixture at the rear end of the tube is used at the close of the operation for aspirating. The ammonia is absorbed and titrated in the usual way.

The New Jersey Agricultural Experiment Station (Rep. N. J. Expt. Sta., 1887, p. 169) has determined the nitrogen in 140 samples of complete fertilizers, 22 samples of nitrogenous matter of high grade, and 19 samples of ground bone, both by the soda-lime method and the Kjeldahl method. The latter gave on the average .04 per cent. more nitrogen in complete fertilizers, .05 per cent. more in high-grade nitrogenous matter, and .10 per cent. in bone.

Reference should also be made to an elaborate series of experiments and observations on the soda-lime method made in the laboratory of the Wesleyan University by or under the direction of Prof. W. O. Atwater (Am. Chem. Jour. 9, 311; 10, 111, 113, 197, 262). A complete abstract of these papers is not deemed necessary here. Professor Atwater says, in closing his paper: ".The perfection to which Kjeldahl's method has lately been brought, and its accuracy, convenience, and inexpensiveness, have led us to its use in this as in many other laboratories. Our experience leads us to decidedly prefer it to the soda-lime method, though we find it advantageous to use both, making one check the other. But the danger of incomplete ammonification of some classes of compounds, *e. g.*, alkaloids, makes us feel it necessary to control both by the absolute method for all classes of substances except those for which they have been most thoroughly tested."

RESULTS OF WORK DONE.

Last November the chairman of your committee sent out to thirty-one official chemists and others five samples for nitrogen determinations. Twelve have reported results.

Sample No. 1, for the purpose of having the various chemists test the accuracy of their standard solutions, pipettes, burettes, and manipulation.

Samples No. 2 and 5, for comparison of soda-lime method with Kjeldahl.

Samples 1, 3, and 4, to compare the Kjeldahl modified, the Ruffle, and the absolute methods.

In sample No. 4, a chloride was put in for the purpose of seeing whether by the Kjeldahl modified method loss of nitrogen would not follow upon putting sulphuric acid on the substance.

COMPOSITION OF THE SAMPLES.

	Per cent. nitrogen.
No. 1. Potassium nitrate, C. P. (dried at 100 degrees)	13.85
No. 2. Cotton-seed meal.	
No. 3. Sodium nitrate, C. P. (dried at 100 degrees)	8.00
Ammonium sulphate, C. P. (dried at 100 degrees)	6.20
Cotton-seed meal	17.00
Acid phosphate	68.80
Theoretical per cent. of nitrogen	3.93

Per cent. nitrogen.

No. 4. Sodium nitrate, C. P. (dried at 100 degrees)..................... 10.00
   Cotton-seed meal... 20.00
   Muriate of potash.. 10.00
   Acid phosphate... 60.00
  Theoretical per cent. of nitrogen 3.16
No. 5. A mixed tankage of the trade.
No. 6. Sample of commercial fertilizer sent by Dr. E. H. Jenkins, Connecticut
   Experiment Station.
No. 7. Pure sodium nitrate 16.47

All the materials, except the tankage, were sifted through a 40-mesh sieve and well mixed before weighing out. The tankage was sifted through a 20-mesh sieve. The constituents, after weighing out, were thoroughly mixed, first with the spatula, on a large sheet of paper, then by running through a drug-mill several times and sifting through the 20-mesh sieve. The mixture was then bottled and sealed as quickly as possible.

A determination by Scovell's method on 2.8 grams acid phosphate gave .02 per cent. of nitrogen. A similar experiment on the muriate of potash gave no nitrogen.

RESULTS OBTAINED.

Soda-lime method.

Analyst.	No. 2.	No. 5.
N. W. Lord..	....	2.97
H. W. Wiley..	7.14	2.85
W. J. Gascoyne	7.49	3.01
William Frear......................................	6.62	2.83
Wilkinson (Alabama Experiment Station)...........	7.70	2.84
Michigan Carbon Works (W. L. Snyder)............	7.47	2.99
National Fertilizer Company (J. H. Kelley)........	7.16	2.80
Average..................................	7.26	2.91

Kjeldahl method.

Analyst.	No. 2.	No. 5.
W. J. Gascoyne	7.52	2.98
William Frear......................................	6.63	2.80
H. W. Wiley.......................................	7.50	2.91
E. H. Jenkins......................................	7.59	3.04
A. M. Peter	7.61	3.00
M. A. Scovell......................................	7.61	3.03
H. A. Weber.......................................	7.38	2.97
N. W. Lord..	7.21	2.94
Average.....................................	7.38	2.96

Absolute method.

Analyst.	No. 1.	No. 2.	No. 3.	No. 4.	No. 6.
N. W. Lord	....	7.23	4.06	3.52	....
A. A. Bennett	13.86	....	3.76	....	....
E. H. Jenkins...............	....	....	....	3.21	3.19
Winton	....	....	....	....	3.06
C.*	....	....	....	....	3.38
Average.............	13.86	7.23	4.40	3.36	3.14

* Unknown.

Kjeldahl method modified for nitrates.

Analyst.	No. 1.	No. 2.	No. 3.	No. 4.	No. 5.	No. 6.	No. 7.
W. J. Gascoynô....	13.80	....	3.93	3.13	3.03	....	....
II. W. Wiley........	13.73	....	3.70	3.13	....	....	....
II. A. Weber......	....	....	3.92	3.22	....	....	....
E. II. Jenkins*.....	13.55	....	3.98	3.00	....	3.11	....
William Frear	13.86	....	3.93	2.90	....	....	....
M. A. Scovell	13.77	7.45	3.93	3.18	2.98	3.18	16.41
A. M. Peter........	13.77	7.47	3.96	3.15	3.00	3.17	16.36
G. C. Caldwell	13.86	7.82	3.54	2.48	3.19	....	....
B.†	....	....	....	....	....	3.45	....
C.†	....	....	....	....	....	3.16	....
D.†	....	....	....	....	....	3.16	....
Average	13.76	7.58	3.86	3.03	3.05	3.20	16.39

* Phenol used in place of salicylic acid, and digested one-half hour in the cold before heating.
† Unknown.

Ruffle method.

Analyst.	No. 1.	No. 2.	No. 3.	No. 4.	No. 5.
II. W. Wiley..........................	13.83	7.47	3.86	3.16	2.97
William Frear *	12.04	....	3.48	2.81	....
W. J. Gascoyne.......................	13.08	....	3.02	3.10	....
Wilkinson (Alabama Experiment Station) ...	13.50	7.42	3.87	2.96	3.05
National Fertilizer Company (J.H. Kelley) ...	13.86	7.27	4.00	2.22	2.92
Average........................	13.38	7.38	3.83	2.85	2.98

* Professor Frear reports the following results on Nos. 1, 3, and 4 by using charcoal in the place of sugar: No. 1, 13.78; No. 3, 4.16; No. 4, 3.43.

AVERAGES.

Method.	Sample.	No. of analysts.	Average.	Highest.	Lowest.	Difference.
Soda lime.............{	2	6	7.26	7.70	7.14	0.56
	5	7	2.91	3.01	2.83	0.18
Kjeldahl.............{	2	8	7.38	7.61	6.63	0.98
	5	8	2.96	3.04	2.80	0.24
Kjeldahl modified for nitrates.{	1	7	13.76	13.86	13.55	0.31
	3	8	3.86	3.98	3.54	0.44
	4	8	3.03	3.22	2.48	0.74
Ruffle.................{	1	5	13.38	13.86	12.04	1.82
	3	5	3.83	4.00	3.48	0.52
	4	5	2.85	3.16	2.22	0.94

REMARKS.

In the soda-lime method the results on No. 5 are quite satisfactory, the average being about .1 lower than with the Kjeldahl method.

On sample No. 2, however, the results are very unsatisfactory. Even by leaving out one result so far away from the others as to show an error of the chemist, the results vary too much to draw any other conclusion than that, if the soda-lime method is used, much more care must be taken on such samples as cotton-seed meal. In the Kjeldahl method all results agree closely with two exceptions in No. 2, and one exception in No. 5. Leaving out these exceptions, the results of which are so far out of the way as to reflect inaccuracy on the part of the analyst rather than the method, the results are highly satisfactory.

On No. 2 the average is 7.49, the highest 7.61, the lowest 7.38, and the difference between highest and lowest 0.23.

48

On No. 5, average 3.00, highest 3.04, lowest 2.91, difference 0.13 per cent.

The Kjeldahl shows far better agreeing results than the soda-lime method and gives an average of .23 per cent. higher results on No. 2 and .09 on No. 5.

The Rufile method shows by the results that it is capable of bringing accurate results, but that it seems of difficult manipulation by many chemists.

On sample No. 1, leaving out one result evidently an error, the average is 13.72, the highest 13.86, the lowest 13.50, difference 0.36.

On sample No. 3, the average is 3.91, highest 4.00, lowest 3.86, difference .14.

On sample No. 4, average 2.85, highest 3.16, lowest 2.81, difference 0.35.

The Kjeldahl method modified for nitrates, although a new method and used by the chemists for the first time as an official method, gives very concordant results.

On sample No. 1 the average is 13.76, and leaving out one low result 13.80, theory 13.85, highest 13.86, lowest 13.73, difference 0.13.

On sample No. 3, average 3.91, highest 3.98, lowest 3.70, difference 0.28, calculated result 3.93.

On sample No. 4, average 3.12, highest 3.22, lowest 3.09, difference 0.13, calculated 3.16.

As compared with the Rufile method the modified Kjeldahl shows higher results on No. 1 by .08, on No. 3 the average is identical, on No. 4 the average 0.27 per cent. greater.

The absolute method has been used only by two chemists and the results are at so great a variance that no comparison can be made.

RECOMMENDATIONS.

Your committee would recommend the continuance of the five methods adopted last year, viz:

(1) The absolute method.
(2) The Kjeldahl method.
(3) The Kjeldahl method modified for nitrates.
(4) The Rufile method.
(5) The soda-lime method.

But after a careful consideration of the results obtained above, they would recommend:

(1) That the soda-lime method be given more in detail, in the hopes of obtaining more concordant results.

(2) That detail changes be made in the Kjeldahl methods.

(3) That attention be called to the use of zinc-dust, in the Kjeldahl modified method. Prof. H. A. Weber calls attention to the fact that unless the zinc-dust be in an impalpable powder and well mixed into the solution, the nascent hydrogen fails to come in contact with all the nitro-compounds and therefore reduction is incomplete. As an example he gives the following results:

Sample.	No. 3.	No. 4.
Finely pulverized zinc	3.46	2.80
Zinc impalpable powder	3.92	3.22

Respectfully submitted.

M. A. SCOVELL,
Chairman.
N. T. LUPTON.

Discussion of the report being in order, Messrs. Frear, Gascoyne, and Voorhees stated that they had found trouble with zinc dust, blanks showing some nitrogen.

Dr. Voorhees said that at the New Jersey Station they always made blanks for error, and questioned whether commercial chemists did so.

Professor Scovell said he had found that sulphuric acid often contained nitrogen, but the strictly C. P. from Eimer & Amend was quite free.

Dr. Voorhees then gave an account of the experiments at his station with the Scovell modification of the Kjeldahl method, as published in their last annual report, and described a convenient form of iron gas-pipe condenser. Of the soda-lime method, Dr. Gascoyne said that the necessity for absence of a channel, as shown by Atwater, was an important factor.

Dr. Voorhees pointed to the necessity of fine division of the substance to at least 60 mesh, while coarse samples could be used with the Kjeldahl method.

Professor Scovell called attention to the generally bad results on cotton-seed and blood with soda lime, in the tables given in his report, and a general opinion that a careful understanding of the details of working this method were very necessary, and without this results would easily fall too low.

With the Scovell modification of the Kjeldahl method general satisfaction was expressed by those who had tried it carefully. The report was then adopted.

Professor Stubbs then presented the report of the committee on sugar, as follows:

REPORT OF THE COMMITTEE ON SUGAR.

The increasing interest in sugar-growing in this country is manifested by the large governmental aid recently given to the manufacture of sugar from sorghum and sugar-cane. This interest has created a demand for chemical information relative to all sugar matters, from the raw juice to the manufactured articles, and to meet this demand is one of the increasing duties of this association. A Manual of Sugar Analysis, by J. H. Tucker, published by D. Van Nostrand, New York, in 1881, contains all of the approved methods of analyses up to that date, many of which are still used without modifications in the laboratories of the world. Others have been slightly modified, while a few have been discarded as totally without merit.

In all sugars and sugar products, including raw juices, the following determinations are needed for industrial work:

(1) Total solids.
(2) Percentage of sucrose.
(3) Percentage of invert sugars.
(4) Ash.

In raw juices.—(1) Syrups and molasses: The total solids may be obtained in the following manner:

(1) By accurately graduated spindles, preferably Baix.
(2) By taking specific gravity.
(3) By evaporating to dryness a weighed quantity mixed with ignited sand in a tarred dish.
(4) A method (Wiley's) of drying and weighing a piece of filter paper, say 12 by 2 inches, saturating with juice or with molasses (easily disseminated over the paper by means of a few drops of alcohol), drying and reweighing.

7717—No. 19——4

Of the above methods only the third is absolutely correct. The fourth in our hands gives results slightly low, while second and first are only approximate. However, the first method, when accurately graduated spindles are used, corrections made for temperature are sufficiently approximate for industrial work. A sufficient time should elapse between the pressing of the cane and the reading of the spindle to permit all the air bubbles to escape; one-half hour is time enough.

Sucrose.—For sugars, raw juices, and sirups sucrose is best determined by a simple polariscopic test, with the following precautions:

(1) Proper sampling of substance.
(2) Careful weighing of sample.
(3) Avoid excess of subacetate of lead.
(4) Dissipate bubbles, which prevent accurate gauging, by a drop or two of ether.
(5) Perfect filling of polariscope tube.
(6) Careful reading of polariscope.
(7) Making all tests in duplicate.
(8) Testing the accuracy of polariscope daily. Instead of direct weighing of substance the Ventzke's process may be used. Following the above (see Tucker, pages 261 and 262), all ripe sorghum juices, all sugars, and sirups of above grade give accurate results. With immature sorghum juices double polarizations show increased results. (See Table No. 1.)

TABLE No. 1.—*Analyses of varieties of sorghum from experimental plats of sugar experiment station in Louisiana, made July 12 and 13; all immature.*

Number of variety from experimental plats.	Total solids by Bux.	Per cent. sucrose, polarization.		Per cent. glucose.
		Single.	Double.	
1	6.8	1.0	1.81	3.40
2	7.0	2.0	2.96	1.90
3	9.8	4.8	5.78	1.50
4	6.5	0.4	2.00	3.29
5	10.9	5.4	6.90	1.82
6	9.0	2.3	3.95	2.12
7	11.7	6.0	7.50	2.13
8	10.6	4.8	6.67	2.68
9	13.3	6.9	8.81	4.25
10	8.5	2.0	4.18	3.40
11	11.1	6.2	8.89	1.70
12	8.9	0.0	1.53	6.34
14	13.2	7.0	7.78	3.71
15	13.6	8.4	9.20	2.75
16	13.3	8.3	8.95	2.85
17	11.8	4.0	5.12	3.40
18	11.5	5.2	6.22	3.30
19	9.8	2.2	3.22	2.95

With molasses of all kinds, at least from Louisiana cane, simple polarization gives results always too low, especially if during the process of manufacture much sugar has been inverted. Double polarization by Clerget's method of inversion is then usually used for determining the sucrose in these bodies, and when invert sugar is the only optically active body, is approximately accurate. When optically active bodies other than invert sugar are present, recourse must be had to inversion and estimations by Fehling's or Violette's methods. Tuchsmidt has confirmed the accuracy of Clerget's process, and Reichardt and Bellman have perfected it, giving us what is now known to us as Clerget's method.

Creydt, on the other hand, has corrected the constant 144 of Clerget and substituted therefore 142. He further prescribes HCl of 1.1888 specific gravity, equal to 38 per cent. of acid, and washes with this acid the bone charcoal used for filter, dries, and pulverizes it. Herzfeld thinks the exact truth as to this constant is not yet known on account of the difficulties, first, of maintaining the temperature of observation ; second, the nature of the so-called "pure sugar" used. Landolt recently adds other errors, among others, that 1° Ventzke (white light)$=0.34455$ circular degrees (sodium light) is too high, and that diverse active substances give numbers sensibly different. He has found with Dr. Rallyan for 1° Ventzke (white light). Sucrose $=$ 0.3465, dextrose $= 0.3448$, invert sugar $= 0.3433$ circular degrees. Landolt finds the constant of Clerget to be 142.4, an expression believed to-day to be the most exact. Herzfeld gives a perfected Clerget method which has the following differences: Water of local temperature surrounds the tube, which cools the solution, at once permitting accurate reading of temperature. The use of washed and dried bone-black after Reichart and Bellman, and calculation of results according to the formula of Clerget, modified by Landolt, viz:

$$R = \frac{100S}{142.4 - \frac{1}{4}T}$$

The following table of work, carefully done by my assistant, Prof. W. L. Hutchinson, with molasses from different plantations in Louisiana, shows results by double polarization and inversion and estimation with Fehling's solution.

TABLE No. 2.

Number.	Where from—plantation.	Beaumé.	Specific grav. ity.	Total solids.	Per cent. sucrose by double polariscope.	Per cent glucose.	Other solids.	Per cent. sucrose by Fehling.	Total sugar as glucose.
1	Belle Chasce....................	41.4	1.4039	78.2	39.59	24.32	15.03	39.65	65.06
2	John Crossly & Sons.........	41.8	1.4092	79.0	41.68	22.69	14.44	40.61	65.44
3	Joseph Garr	40.0	1.3933	76.6	34.91	29.19	9.05	32.25	63.14
4	Boujere	40.7	1.3940	76.7	37.62	24.77	15.32	37.53	61.28
5	Belle Terre..................	41.6	1.4005	78.6	42.57	22.78	14.63	41.30	66.26
6	Woodlawn	42.3	1.4165	80.1	33.42	33.75	14.49	32.79	68.22
7	Crescent	42.3	1.4165	80.1	36.44	28.42	14.93	37.67	67.08
8	Mary.......................	42.3	1.4165	80.1	38.82	26.86	16.72	30.08	67.84
9	Perseverance	43.2	1.4287	81.9	40.31	20.00	22.93	43.54	65.84
10	David......................	42.3	1.4165	80.1	37.62	26.92	16.72	36.97	65.84
11	Orange Grove..............	42.4	1.4179	80.3	42.25	26.73	13.05	39.94	68.78
12	Palo Alto..................	42.6	1.4205	80.7	38.14	25.96	16.43	40.28	68.36
13	Hester	42.5	1.4192	80.5	34.72	30.17	18.52	33.14	65.06

While the concurrence of results in the main are satisfactory, there are some which could not be made to agree, suggesting strongly either a fault in the method or other optically active bodies than invert sugar, perhaps both; and while for the present we recommend Clerget's method we earnestly request an investigation of Herzfeld's modification of Clerget's method during the coming year by all sugar chemists.

INVERT SUGAR.

The exact estimation of invert sugar in sugar matters has been recently ably investigated and discussed. Heretofore invert sugar has been determined by a simple titration with the cupric tartrate liquor of Fehling, Violette, etc. The formulas for the preparation of these solutions are numerous. They all contain sulphate of copper. rochelle salts, and caustic alkali and present little real difference. Meissl and Herz, feld have described a process based upon the weight of copper reduced from the Fehling solution when accompanied by particular conditions. In spite of its great precision it has found only limited application, especially in France, where its numerous difficulties have prevented its use on an extensive scale. Beggart employs Fehling's solution, which he keeps in bottles (bouchées à l'émeri) and which are placed in a dark place. To titrate this liquid they dissolve 0.95 gram of pure sugar in a little water with 2 cubic centimeters of hydrochloric acid, and after adding 500 cubic centimeters of water it is gently heated for ten minutes at 70° C. This solution is cooled, and saturated with bicarbonate of soda and raised to one liter; .25 cubic centimeters of this liquid will reduce 5 cubic centimeters of Fehling's solution, the end reaction being determined as usual by means of ferrocyanide of potash and acetic acid, using Wiley's tubes.

To preserve a standard solution of invert sugar he prepares a 15 per cent. solution and make strongly alkaline. In this way it will be preserved many months. With this solution he daily tests his Fehling's solution. This constitutes the only merit of this process, a merit which should be observed in every process using the cupric tartrate process.

Patterson's process is based upon another principle. To 100 cubic centimeters of Fehling's solution he adds a quantity of sugar insufficient to precipitate all the copper, and determines excess of copper by a titration with a normal invert sugar solution containing .002 grams per cubic centimeter. Herzfeld rightly condemns the process, since no correct conclusions from the use of the cupric tartrate solution can be drawn without considering the strength of ebullition and influence of crystallizable sugar upon the reducing power of invert sugar.

All the copper liquids so far proposed, although different in composition, have one common property, viz, they all contain a salt of copper, a double tartrate of potash and soda, and caustic or carbonated alkali. They are all reduced not only by invert sugar but also by dextrine and sometimes even by sucrose. After a certain time they undergo alteration, deposit the oxide of copper, especially if exposed to light.

Soldaini's solution does not possess these objectionable qualities, at least so says Degener, who has studied carefully its properties.

This liquid is prepared, according to Degener, in following manner: (1) 40 grams of sulphate of copper is dissolved in water, and, in another vessel, 40 grams of carbonate of soda is also dissolved in water. The two solutions are mixed and the copper precipitated in the state of hydrobasic carbonate according to following equation: $2 Cu SO_4 + 2Na_2CO_3 + H_2O = CuCO_3CuOH_2O + CO_2 + 2 Na_2SO_4$. The precipitate is washed with cold water and dried. This precipitate (15 grams) is added to a very concentrated and boiling solution of bicarbonate of potash (about 415 grams) and agitated until the whole is completely, or nearly so, dissolved, water is added to form a volume of 1,400 cubic centimeters, and the whole mass heated for two hours upon a water bath. The insoluble matter is filtered and the filtrate, after cooling, is of a deep blue color and has a density of about 22°.5 Baumé (1.185). The sensibility of this liquid is so great that it gives a decided reaction with .0014 grams of invert sugar. The presence of sucrose in the solution increases still this sensibility. Ammoniacal salts are not hurtful to the reaction, but they do sometimes retain in solution a slight quantity of copper.

It is then indispensable to boil the solution for five minutes at least in order to drive off ammonia. The quantitative test for sugar is made in the following manner: Fifty cubic centimeters of Soldaini's solution is boiled for five minutes upon a water bath, and 15 cubic centimeters of the sugar liquid (containing 100 grams of matter in 100 cubic centimeters of water) added, and boiled again for five minutes. After a rapid cooling the liquid is thrown upon a Swedish filter and washed with water until every trace of a blue color disappears. Then the paper is removed and the precipitate examined, which should be of a clear red color.

Bodenbacnder and Schiller (Zeitschrift des Vereins für Rübenzucker-Industrie, 1887, page 135) have determined the following quantitative process with this liquid: One hundred to 150 cubic centimeters of Soldaini's solution is placed in an Erlenmeyer flask and heated for five minutes over a gas jet. A solution containing 10 grams of sugar (clarified with subacetate of lead if necessary) is added, and heated again for five minutes, always over a direct flame. This precipitates all the copper, to which is added 100 cubic centimeters of distilled water, in order to cool the mass very quickly. The trouble dliquid is filtered (preferably through a Sohxlet or Herzfield filter) and washed with distilled water. This oxide of copper is reduced in a current of hydrogen, and the copper is weighed in a metallic form, and this weight multiplied by .3546 gives the weight of invert sugar. By this method .01 per cent. of invert sugar can be accurately determined.

Sidersky has recently offered a new volumetric method based upon the use of Soldaini's solution. With sugars the same method as is now in use with Fehling's solution can easily be followed, watching the disappearance of the blue color, and testing the end with ferrocyanide of potash and acetic acid. This process offers no serious objections common to Fehling's solution, but is inapplicable to colored sugar solutions, such as molasses, etc. For the last the following is recommended: twenty-five grams of molasses is dissolved in 100 cubic centimeters of water and subacetate of lead added in sufficient quantities to precipitate the impurities, and the volume raised to 200 cubic centimeters and filtered. To 100 cubic centimeters of the filtrate are added 25 cubic centimeters of concentrated solution of carbonate of soda, agitated and filtered again. One hundred cubic centimeters of the second filtrate with excess of lead removed is taken for analysis. On the other hand, 100 cubic centimeters of Soldaini's solution is added to a flask of Bohemian glass and heated five minutes over an open flame. The sugar solution is now added a little by little, and the heating continued for five minutes. Finally, the heat is withdrawn and cooled by turning in 100 cubic centimeters of cold water, and filtered through a Swedish filter, washed with hot water, letting each washing run off before another addition. Three or four washings will generally remove completely the alkaline reaction. The precipitate is then washed through a hole in the filter into a flask, removing the last trace of copper. Twenty-five cubic centimeters of normal sulphuric acid is added, with two or three crystals of chlorate of potash, and the whole gently heated to dissolve completely the oxide of copper, which is transformed into copper sulphate, according to equation, $3Cu_2O + 6H_2SO_4 + KClO_3 = 6CuSO_4 + 6H_2O + KCl$. The excess of sulphuric acid is determined by a standard ammonia solution (semi-normal) of which the best indicator is the sulphate of copper itself. When the deep blue color gives place to a greenish tinge the titration is completed. The method of titration is performed as follows: Having cooled the contents of the flask, a quantity of ammonia equivalent to 25 cubic centimeters of normal sulphuric acid added. From a burette graduated into one-tenth cubic centimeters standard sulphuric acid is dropped in drop by drop, agitating after each addition. The blue color disappears with each addition to reappear after shaking. When the last trace of ammonia is saturated the titration is complete, which is known by a very feeble greenish tinge. The number of cubic centimeters is read from the burette, which is equivalent to the copper precipitated. The equivalent of copper is 31.7, and the normal acid equivalent is .0317 of copper. Multiplying the copper found by 3546 we have the invert sugar. A blank titration is needed to accurately determine the slight excess which gives the pale green tinge. This process is to be highly recommended if experiments show it to be as accurate in our hands as it has in France.

(4) *Ash.*—For the determination of ash, Scheibler's process of ignition with H_2SO_4 and deducting one-tenth is recommended. For details, see Tucker's Manual, page 226.

Your committee, in conclusion, recommends that for technical work the following is admissible:

(1) Taking the total solids with an accurately graduating Brix spindle and correcting for temperature.

(2) A polariscopic test for raw juices, low sirups and sugars, using the precautions mentioned in this report.

(3) For heavy sirups, molasses, and tank bottoms, double polarization with Clerget's tables.

(4) For invert sugar, use Fehling's or Violett's solutions, preferably the last, using as an end reaction ferrocyanide of potash and acetic acid; best accomplished by use of Wiley's tubes.

(5) Ignition with sulphuric acid at a low red heat and deducting one-tenth for ash. For laboratory work the above may be substituted by some of the more accurate methods given above.

Respectfully submitted.

W. C. STUBBS,
Chairman.

REMARKS ON PROFESSOR STUBBS'S REPORT.

By H. W. WILEY.

Mr. PRESIDENT. In connection with the report of Professor Stubbs on sugar analysis, which has just been presented to the association, I beg leave to call the attention of the members to a standard polariscope which I have had constructed by Schmidt & Haensch in Berlin. This instrument is the double compensating apparatus manufactured by that firm, which has already gained great favor among chemists. The apparatus which I show you was specially constructed for the use of this Department, and is capable of receiving an observation tube 600 millimeters in length.

By means of the double compensating apparatus any error which may be due to an inaccuracy in the quartz wedge is reduced to a minimum. The apparatus has the advantage of using ordinary lamp-light, which is of great importance when outside-station work is to be considered, where the difficulty of securing a monochromatic flame is great. This instrument has lately been examined by Professor Andrews, the expert employed by the United States Treasury to investigate the methods of sugar analysis employed in the United States custom-houses. He proved it to be as nearly exact as any instrument can be made. It is not necessary to mention that for general purposes of research a rotation instrument like the large model Laurent is preferable to any form of compensating apparatus; but in sugar work, where an arbitrary scale is used, I think an instrument of this kind would be found superior to anything now in use.

The report of the committee on sugar was then adopted. The committee on amendments to the constitution then submitted the following:

Amendment to article 4.

Article 4 to read as follows:

There shall be appointed by the president at the regular annual meeting a reporter for each of the subjects to be considered by the association.

It shall be the duty of these reporters to test methods, to prepare and distribute samples and standard re-agents to members of the association and others desiring the same; to furnish blanks for tabulating analyses, and to present at the annual meeting the results of work done, discussion thereof and recommendations of methods to be followed.

Addition to article 2.

All members of the association who lose their right to such membership by retiring from positions indicated as requisite for membership shall be entitled to become honorary members and to all privileges of membership save the right to hold office and to vote.

Approved.

H. W. WILEY.
WM. C. STUBBS.
E. B. VOORHEES.

The report of the committee was then adopted, and the amendments made part of the constitution.

The committee on elections then reported that "hereafter all officers be elected by ballot."

In accordance therewith the convention proceeded to an election, with the following result:

President: Prof. John A. Myers, of West Virginia.

Vice-president: Prof. M. A. Scovell, of Kentucky.

Secretary: Mr. Clifford Richardson, of Washington.

Executive committee: Dr. H. W. Wiley, of Washington; Dr. William Frear, of Pennsylvania.

It was then agreed that the committee on branding bags, labels, etc., be continued, with Professor White substituted for Professor McMurtrie.

Professor Myers then moved, and it was adopted:

That the hearty thanks of the convention be returned to Hon. Norman J. Colman, Commissioner of Agriculture, for his courtesies and cordial co-operation in the work of the association.

The convention then, on motion of Professor Scovell, adjourned *sine die.*

METHODS FOR DETERMINING PHOSPHORIC ACID AND MOISTURE.

(1) *Preparation of sample.*—The sample should be well intermixed and properly prepared, so that separate portions shall accurately represent the substance under examination, without loss or gain of moisture.

(2) *Determination of moisture.*—(*a*) In potash salts, nitrate of soda, and sulphate of ammonia heat 1 to 5 grams at 130° C. till the weight is constant, and reckon water from the loss. (*b*) In all other fertilizers heat 2 grams, or if the sample is too coarse to secure uniform lots of 2 grams each, 5 grams, for five hours, at 100° in a steam bath.

(3) *Water-soluble phosphoric acid.*—Weigh out 2 grams in a small beaker, wash by decantation four or five times with not more than from 10 to 15 cubic centimeters of water, then rub it up in the beaker with a rubber-tipped pestle to a homeogeneous paste, and then wash four or five times by decantation with from 10 to 15 cubic centimeters of water. Transfer the residue to a 9 cubic centimeter, No. 589 Schleicher and Schüll filter, and wash with water until the filtrate measures not less than 250 cubic centimeters. Mix the washings. Take an aliquot (corresponding to $\frac{1}{2}$ gram) and determine phosphoric acid, as under total phosphoric acid.

(4) *Citrate-insoluble phosphoric acid.*—Wash the residue of the treatment with water into a 200 cubic centimeter flask with 100 cubic centimeters of strictly neutral ammonium citrate solution of 1.09 density, prepared as hereafter directed. Cork the flask securely and place it in a water bath, the water of which stands at 65° C. (The water bath should be of such a size that the introduction of the cold flask or flasks shall not cause a reduction of the temperature of the bath of more than 2° C.) Raising the temperature as rapidly as practicable to 65° C., which is subsequently maintained, digest with frequent shakings for thirty minutes from the instant of insertion, filter the warm solution quickly (best with filter-pump), and wash with water of ordinary temperature. Transfer the filter and its contents to a capsule, ignite until the organic matter is destroyed, treat with 10 to 15 cubic centimeters

57

of concentrated hydrochloric or nitric acid, digest over a low flame until the phosphate is dissolved, dilute to 200 cubic centimeters, mix, pass through a dry filter, take an aliquot and determine phosphoric acid as under total.

In case a determination of citrate-insoluble phosphoric acid is required in non-acidulated goods, it is to be made by treating 2 grams of the phosphatic material, without previous washing with water, precisely in the way above described, except that in case the substance contains much animal matter (bone, fish, etc.), the residue insoluble in ammonium citrate is to be treated by one of the processes described below.

(5) *Total phosphoric acid.*—Weigh 2 grams and treat by one of the following methods: (1) Evaporation with 5 cubic centimeters magnesium nitrate, ignition, and solution in acid. (2) Solution in 30 cubic centimeters concentrated nitric acid with a small quantity of hydrochloric acid. (3) Add 30 cubic centimeters concentrated hydrochloric acid, heat, and add cautiously and in small quantities at a time about 0.5 gram of finely pulverized potassium chlorate.

Boil gently until all phosphates are dissolved and all organic matter destroyed; dilute to 200 cubic centimeters; mix and pass through a dry filter; take 50 cubic centimeters of filtrate; neutralize with ammonia (in case hydrochloric acid has been used as a solvent add about 15 grams dry ammonium nitrate or its equivalent). To the hot solutions for every decigram of P_2O_5 that is present, add 50 cubic centimeters of molybdic solution. Digest at about 65° C. for one hour, filter, and wash with water or ammonium nitrate solution. (Test the filtrate by renewed digestion and addition of more molybdic solution.) Dissolve the precipitate on the filter with ammonia and hot water and wash into a beaker to a bulk of not more than 100 cubic centimeters. Nearly neutralize with hydrochloric acid, cool, and add magnesia mixture from a burette; add slowly (one drop per second), stirring vigorously. After fifteen minutes add 30 cubic centimeters of ammonia solution of density 0.95. Let stand several hours (two hours is usually enough). Filter, wash with dilute ammonia, ignite intensely for ten minutes, and weigh.

(6) Citrate-soluble phosphoric acid. The sum of the water-soluble and citric insoluble subtracted from the total gives the citrate-soluble.

PREPARATION OF REAGENTS.

(1) *To prepare ammonium citrate solution.*—Mix 370 grams of commercial citric acid with 1,500 cubic centimeters of water; nearly neutralize with crushed commercial carbonate of ammonia; heat to expel the carbonic acid; cool; add ammonia until exactly neutral (testing by saturated alcoholic solution of coralline) and bring to volume of two liters. Test the gravity, which should be 1.09 at 20° C., before using.

(2) *To prepare molybdic solution.*—Dissolve 100 grams of molybdic acid in 400 grams or 417 cubic centimeters of ammonia of specific grav-

ity 0.96, and pour the solution thus obtained into 1,500 grams or 1,250 cubic centimeters of nitric acid of specific gravity 1.20. Keep the mixture in a warm place for several days, or until a portion heated to 40° C. deposits no yellow precipitate of ammonium phospho-molybdate. Decant the solution from any sediment, and preserve in glass-stoppered vessels.

(3) *To prepare ammonium nitrate solution.*—Dissolve 200 grams of commercial ammonium nitrate in water and bring to a volume of two liters.

(4) *To prepare magnesia mixture.*—Dissolve 22 grams of recently ignited calcined magnesia in dilute hydrochloric acid, avoiding excess of the latter. Add a little calcined magnesia in excess, and boil a few minutes to precipitate iron, alumina, and phosphoric acid; filter, add 280 grams of ammonium chloride, 700 cubic centimeters of ammonia of specific gravity 0.96, and water enough to make the volume of two liters. Instead of the solution of 22 grams of calcined magnesia 110 grams of crystallized magnesium chloride ($MgCl_2$, $6H_2O$) may be used.

(5) *Dilute ammonia for washing.*—One volume ammonia of specific gravity 0.96 mixed with three volumes of water, or usually one volume of concentrated ammonia with six volumes of water.

(6) *Nitrate of magnesia.*—Dissolve 320 grams of calcined magnesia in nitric acid, avoiding an excess of the latter; then add a little calcined magnesia in excess; boil; filter from excess of magnesia, ferric oxide, etc., and bring to volume of two liters.

METHODS OF DETERMINING POTASH.

METHOD OF LINDO AS MODIFIED BY GLADDING.

(1) *Superphosphates.*—Boil 10 grams of the fertilizer with 300 cubic centimeters of water for ten minutes. Cool the solution; add a little oxalate of ammonia and then ammonia in slight excess, thus precipitating all phosphate and sulphate of lime, oxide of iron, and alumina, etc.; make up to 500 cubic centimeters, mix thoroughly and filter through a dry filter; take 50 cubic centimeters corresponding to 1 gram, evaporate nearly to dryness, add 1 cubic centimeter of dilute H_2SO_4 (1 to 1), and evaporate to dryness and ignite to whiteness. As all the potash is in form of sulphate, no loss need be apprehended by volatilization of potash, and a full red heat must be used until the residue is perfectly white. This residue is dissolved in hot water plus a few drops of HCl; 5 cubic centimeters of a solution of pure NaCl (containing 20 grams NaCl to the liter) and an excess of platinum solution (4 cubic centimeters) are now added. This solution is then evaporated to dryness in a small dish, the residue taken up with a little water, sufficient to dissolve it, and strong alcohol added. The precipitate is washed thoroughly with alcohol by decantation and on filter as usual. The wash-

ing should be continued even after the filtrate is colorless. Ten cubic centimeters of the NH_4Cl solution prepared as directed are now run through the filter, or the washing may be performed in the dish. These 10 cubic centimeters will contain the bulk of the impurities, and are thrown away. Fresh portions of 10 cubic centimeters NH_4Cl are now run through the filter several times (five or six). The filter is then washed thoroughly with pure alcohol, dried, and weighed as usual. The platinum solution used contains 1 gram metallic platinum in every 10 cubic centimeters.

(2) *Muriates of potash.*—In the analysis of these salts an aliquot portion, containing .500 gram, is evaporated with 10 cubic centimeters platinum solution plus a few drops of HCl, and washed as before.

(3) *Sulphate of potash, kainite, etc.*—In the analysis of these salts an aliquot portion containing .500 gram is taken, .250 gram of $NaCl$ added, plus a few drops of HCl, and the whole evaporated with 15 cubic centimeters platinum solution. In this case special care must be taken, in the washing with alcohol, to remove all the double chloride of platinum and sodium. The washing should be continued for some time after the filtrate is colorless. Twenty-five cubic centimeters of the NH_4Cl solution are employed, instead of 10 cubic centimeters, and the 25 cubic centimeters poured through at least six times to remove all sulphates and chlorides. Wash finally with alcohol, dry, and weigh as usual.

To prepare the washing solution of NH_4Cl, place in a bottle 500 cubic centimeters H_2O, 100 grams of NH_4Cl; shake till dissolved. Now pulverize 5 or 10 grams of K_2PtCl_6, put in the bottle, and shake at intervals for six or eight hours; let settle over night; then filter off liquid into a second bottle. The first bottle is then ready for a preparation of a fresh supply when needed.

ALTERNATE METHOD.

In case the potash is contained in organic compounds, like tobacco stems, cotton-seed hulls, etc., it is to be saturated with strong sulphuric acid and ignited in a muffle to destroy organic matter. Pulverize the fertilizer (200 or 300 grams) in a mortar; take 10 grams, boil for ten minutes with 200 cubic centimeters water, and after cooling, and without filtering, make up to 1,000 cubic centimeters, and filter through a dry paper. If the sample have 10 to 15 per cent. K_2O (kainite), take 50 cubic centimeters of the filtrate; if from 2 to 3 per cent. K_2O (ordinary potash fertilizers), take 100 cubic centimeters of the filtrate. In each case make the volume up to 150 cubic centimeters, heat to 100°, and add, drop by drop, with constant stirring, slight excess of barium chloride; without filtering, in the same manner, add barium hydrate in slight excess. Heat, filter, and wash until precipitate is free of chlorides. Add to filtrate 1 cubic centimeter strong ammonium hydrate, and then a saturated solution of ammonium carbonate until excess of barium is precipitated. Heat. Add now, in fine powder, 0.5 gram pure oxalic acid or 0.75 gram ammonium oxalate. Filter, wash free of chlorides, evaporate filtrate to dryness in a platinum dish, and, holding dish with crucible tongs, ignite carefully over the free flame below red heat until all volatile matter is driven off.

The residue is now digested with hot water, filtered through a small filter, and washed with successive small portions of water until the filtrate amounts to 30 cubic centimeters or more. To this filtrate, after adding two drops of strong hydrochloric acid, is added, in a porcelain dish, 5 to 10 cubic centimeters of a solution of 10 grams of platinic chloride in 100 cubic centimeters of water. The mixture is now evaporated on the water-bath to a thick sirup, or further, as above treated with strong alcohol, washed by decantation, collected in a Gooch crucible or other form of filter, washed with strong alcohol, afterwards with 5 cubic centimeters ether, dried for thirty minutes at 100° C., and weighed.

It is desirable, if there is an appearance of foreign matter in the double salt, that it should be washed, according to the previous method, with 10 cubic centimeters of the half-concentrated solution of NH_4Cl, which has been saturated by shaking with K_2PtCl_6, as recommended by Gladding.

The use of the factor 0.3056 for converting K_2PtCl_6 to KCl and 0.19308 for converting to K_2O are continued.

METHODS FOR THE DETERMINATION OF NITROGEN.

THE ABSOLUTE OR CUPRIC OXIDE METHOD.

(Applicable to all nitrogen determinations.)

The apparatus and re-agents needed are as follows:

APPARATUS.

Combustion tube of best hard Bohemian glass, about 26 inches long and one-half inch internal diameter.

Azotometer of at least 100 cubic centimeters capacity, accurately calibrated.

Sprengel mercury air pump.

Small paper scoop, easily made from stiff writing-paper.

RE-AGENTS.

Cupric oxide (coarse).—Wire form; to be ignited and cooled before using.

Fine cupric oxide.—Prepared by pounding ordinary cupric oxide in mortar.

Metallic copper.—Granulated copper or fine copper gauze reduced and cooled in stream of hydrogen.

Sodium bicarbonate.—Free from organic matter.

Caustic potash solution.—Dissolve commercial stick potash in less than its weight of water so that crystals are deposited on cooling. When absorption of carbonic acid ceases to be prompt solution must be discarded.

LOADING TUBE.

Of ordinary commercial fertilizers take 1 to 2 grams for analysis. In the case of highly nitrogenous substances the amount to be taken must be regulated by the amount of nitrogen estimated to be present. Fill tube as follows: (1) About 2 inches of coarse cupric oxide. (2) Place on the small paper scoop enough of the fine cupric oxide to fill, after having been mixed with the substance to be analyzed, about 4 inches of the tube; pour on this the substance, rinsing watch glass with a little of the fine oxide, and mix thoroughly with spatula; pour into tube, rinsing the scoop with a little fine oxide. (3) About 12 inches of coarse cupric oxide. (4) About 3 inches of metallic copper. (5) About 2½ inches of coarse cupric oxide (anterior layer). (6) Small plug of asbestos. (7) Eight-tenths to 1 gram of sodium bicarbonate. (8) Large, loose plug of asbestos; place tube in furnace, leaving about 1 inch of it pro- jecting; connect with pump by rubber stopper smeared with glycerine, taking care to make connection perfectly tight.

OPERATION.

Exhaust air from tube by means of pump. When a vacuum has been obtained allow flow of mercury to continue, light gas under that part of tube containing metallic copper, anterior layer of cupric oxide (see 5th above) and bicarbonate of soda. As soon as vacuum is destroyed and apparatus filled with carbonic acid gas, shut off the flow of mer- cury and at once introduce the delivery tube of the pump into the re- ceiving arm of the azotometer, and just below the surface of the mercury seal of the azotometer, so that the escaping bubbles will pass into the air and not into the azotometer, thus avoiding the useless saturation of the caustic potash solution.

When the flow of carbonic acid has very nearly or completely ceased, pass the delivery tube down into the receiving arm, so that the bubbles will escape into the azotometer. Light the jets under the 12-inch layer of oxide, heat gently for a few moments to drive out any moisture that may be present, and bring to red heat. Heat gradually mixture of substance and oxide, lighting one jet at a time. Avoid too rapid evolution of bub- bles, which should be allowed to escape at rate of about one per second or a little faster.

When the jets under mixture have all been turned on, light jets under layer of oxide at end of tube. When evolution of gas has ceased turn out all the lights except those under the metallic copper and anterior layer of oxide, and allow to cool for a few moments. Exhaust with pump and remove azotometer before flow of mercury is stopped. Break con-

nection of tube with pump, stop flow of mercury, and extinguish lights
Allow azotometer to stand for at least an hour or cool with stream of
water until permanent volume and temperature are reached.

Adjust accurately the level of the KOH solution in bulb to that in
azotometer, note volume of gas, temperature, and height of barometer;
make calculations as usual. The labor of calculation may be much
diminished by the use of the tables prepared by Messrs. Battle and Dancy,
of the North Carolina Experiment Station (Raleigh, N. C.).

The above details are, with some modifications, those given in the report of the Connecticut Station for 1879 (p. 124), which may be consulted
for details of apparatus, should such details be desired.

THE KJELDAHL METHOD.

(Not applicable in presence of nitrates.)

APPARATUS AND REAGENTS.

(1) Hydrochloric acid, whose absolute strength has been determined
(a) by precipitating with silver nitrate and weighing the silver chloride,
(b) by sodium carbonate, as described in Fresenius's Quantitative
Analysis, second American edition, page 680, and (c) by determining
the amount neutralized by the distillate from a weighed quantity of
pure ammonium chloride boiled with an excess of sodium hydrate.
Half normal acid, i. e., containing 18.25 grams hydrochloric acid to the
liter, is recommended.

(2) Standard ammonia whose strength, relative to the acid, has been
accurately determined. One-tenth normal ammonia, i. e., containing
1.7 grams ammonia to the liter, is recommended for accurate work.

(3) "C. P." sulphuric acid, specific gravity 1.83, free from nitrates
and also from ammonium sulphate, which is sometimes added in the
process of manufacture to destroy oxides of nitrogen. Eimer & Amend's
"Strictly C. P." is good.

(4) Mercuric oxide, HgO, prepared in the wet way. That prepared
from mercury nitrate can not safely be used.

(5) Potassium permanganate tolerably finely pulverized.

(6) Granulated zinc.

(7) A solution of 40 grams of commercial potassium sulphide in one
liter of water.

(8) A saturated solution of sodium hydrate free from nitrates, which
are sometimes added in the process of manufacture to destroy organic
matter and improve the color of the product. That of the Greenbank
Alkali Company is of good quality.

(9) Solution of cochineal prepared according to Fresenius's Quantitative Analysis, second American edition, page 679.

(10) Kjeldahl digestion flasks of hard, moderately thick, well an-
nealed glass. These flasks are about 9 inches long, with a round, pear-
shaped bottom, having a maximum diameter of 2½ inches, and tapering

out gradually in a long neck, which is three-fourths of an inch in diameter at the narrowest part, and flared a little at the edge. The total capacity is 225 to 250 cubic centimeters.

(11) Distillation flasks of ordinary shape, 550 cubic centimeters capacity, and fitted with a rubber stopper and a bulb tube above to prevent the possibility of sodium hydrate being carried over mechanically during distillation. The bulbs are about 1½ inches in diameter, the tubes being the same diameter as the condenser and cut off obliquely at the lower end. This is adjusted to the tube of the condenser by a rubber tube.

(12) *A condenser.*—Several forms have been described, no one of which is equally convenient for all laboratories. The essential thing is that the tube which carries the steam to be condensed shall be of block-tin. The upper ends of the tin tubes should be bent so that the glass connections may have a slope toward the distilling flasks. All kinds of glass are decomposed by steam and ammonia vapor, and will give up alkali enough to impair accuracy. (See Kreussler and Henzold, Ber. Berichte, XVII, 34.) The condenser in use in the laboratory of the Connecticut Experiment Station, devised by Professor Johnson, consists of a copper tank, supported by a wooden frame, so that its bottom is 11 inches above the work-bench on which it stands. This tank is 16 inches high, 32 inches long, and 3 inches wide from front to back, widening above to 6 inches. It is provided with a water supply tube, which goes to the bottom, and a larger overflow pipe above. The block-tin condensing tubes, whose external diameter is three-eighths of an inch, seven in number, enter the tank through holes in the front side of it near the top, above the level of the overflow, and pass down perpendicularly through the tank and out through rubber stoppers tightly fitted into holes in the bottom. They project about 1½ inches below the bottom of the tank, and are connected by short rubber tubes with glass bulb tubes of the usual shape, which dip into precipitating beakers or Erlenmeyer flasks of about 300 cubic centimeters capacity. The titration can be made directly in them. The distillation flasks are supported on a sheet-iron shelf, attached to the wooden frame that supports the tanks in front of the latter. Where each flask is to stand a circular hole is cut, with three projecting lips, which support the wire gauze or asbestus under the flask, and three other lips, which hold the flask in place and prevent its moving laterally out of place while distillation is going on. Below this sheet-iron shelf is a metal tube carrying seven Bunsen burners, each with a stop-cock like those of a gas-combustion furnace. These burners are of a larger diameter at the top, which prevents smoking when covered with fine gauze to prevent the flame from striking back.

(13) The stand for holding the digestion flasks consists of a pan of sheet-iron 29 inches long by 8 inches wide, on the front of which is fastened a shelf of sheet-iron as long as the pan, 5 inches wide and 4 inches

high. In this are cut six holes 1⅜ inches in diameter. At the back of the pan is a stout wire running lengthwise of the stand, 8 inches high, with a bend or depression opposite each hole in the shelf. The digestion flask rests with its lower part over a hole in the shelf and its neck in one of the depressions in the wire frame, which holds it securely in position. Heat is supplied by low Bunsen burners below the shelf. With a little care the naked flame can be applied directly to the flask without danger.

THE DETERMINATION.

(1) *The digestion.*—Seven-tenths to 2.8 grams of the substance to be analyzed, according to its proportion of nitrogen, is brought into a digestion flask with approximately .7 gram of mercuric oxide and 20 cubic centimeters of sulphuric acid. The flask is placed on the frame above described in an inclined position, and heated below the boiling point of the acid for from 5 to 15 minutes, or until frothing has ceased. If the mixture froths badly, a small piece of paraffine may be added to prevent it. The heat is then raised until the acid boils briskly. No further attention is required till the contents of the flask have become a clear liquid, which is colorless, or at least has only a very pale straw color. The flask is then removed from the frame, held upright, and, while still hot, potassium permanganate is dropped in carefully and in small quantity at a time till, after shaking, the liquid remains of a green or purple color.

(2) *The distillation.*—After cooling, the contents of the flask are transferred to the distilling flask with about 200 cubic centimeters of water, and to this a few pieces of granulated zinc and 25 cubic centimeters of potassium sulphide solution are added, shaking the flask to mix its contents. Next add 50 cubic centimeters of the soda solution, or sufficient to make the reaction strongly alkaline, pouring it down the side of the flask so that it does not mix at once with the acid solution. Connect the flask with the condenser, mix the contents by shaking, and distill until all ammonia has passed over into the standard acid. The first 150 cubic centimeters of the distillate will generally contain all of the ammonia. This operation usually requires from forty minutes to one hour and a half. The distillate is then titrated with standard ammonia.

The use of mercuric oxide in this operation greatly shortens the time necessary for digestion, which is rarely over an hour and a half in cases of substances most difficult to oxidize, and is more commonly less than an hour. In most cases the use of potassium permanganate is quite unnecessary, but it is believed that in exceptional cases it is required for complete oxidation, and in view of the uncertainty it is always used. Potassium sulphide removes all mercury from solution, and so prevents the formation of mercuro-ammonium compounds which are not completely decomposed by soda solution. The addition of zinc gives rise to an evolution of hydrogen and prevents violent bumping. Previous to use the reagents should be tested by a blank experiment with sugar, which will partially reduce any nitrates that are present which might otherwise escape notice.

KJELDAHL METHOD MODIFIED TO INCLUDE THE NITROGEN OF NITRATES.*

(Applicable to all fertilizers containing nitrates.)

Besides the reagents and apparatus given under the Kjeldahl method, there will be needed—

(1) Zinc dust. This should be an impalpable powder; granulated zinc or zinc filings will not answer.

(2) Commercial salicylic acid.

THE DETERMINATION.

Bring from .7 to 1.4 grams of the substance to be analyzed into a Kjeldahl digesting flask, add to this 30 cubic centimeters of sulphuric acid containing 2 grams of salicylic acid, and shake thoroughly; then add *gradually* 3 grams of zinc dust, shaking the contents of the flask at the same time. Finally place the flask on the stand for holding the digestion flasks, where it is heated over a low flame until all danger from frothing has passed. The heat is then raised until the acid boils briskly, and the boiling continued until white fumes no longer pour out of the flask. This requires about five or ten minutes. Add now approximately .7 gram mercuric oxide, and continue the boiling until the liquid in the flask is colorless, or nearly so. (In case the contents of the flask are likely to become solid before this point is reached, add 10 cubic centimeters more of sulphuric acid.) Complete the oxidation with a little permanganate of potash in the usual way, and proceed with the distillation as described in the Kjeldahl method. The reagents should be tested by blank experiments.

THE RUFFLE METHOD.

APPARATUS AND RE-AGENTS.

(1) Standard solutions and indicator the same as for the Kjeldahl method.

(2) A mixture of equal parts by weight of fine soda-lime and finely powdered, crystallized sodium hyposulphite.

(3) A mixture of equal parts by weight of finely powdered granulated sugar and flowers of sulphur.

(4) Granulated soda-lime, as described under the soda-lime method.

(5) Combustion tubes of hard Bohemian glass, 20 inches long and $\frac{1}{2}$ inch in diameter.

(6) Bulbed "U" tubes or Wills's bulbs, as described under the soda-lime method.

* Described by Prof. M. A. Scovell.

PREPARATION.

(1) Clean and fill the U tube with 10 cubic centimeters of standard acid.

(2) Fit cork and glass connecting tube. Fill the tube as follows: (1) A loosely fitting plug of asbestus, previously ignited, and then 1 to 1½ inches of the hyposulphite mixture. (2) The weighed portion of the substance to be analyzed is intimately mixed with from 5 to 10 grams of the sugar and sulphur mixture. (3) Pour on a piece of glazed paper, or porcelain mortar, a sufficient quantity of the hyposulphite mixture to fill about 10 inches of the tube; then add the substance to be analyzed, as previously prepared; mix carefully and pour into the tube; shake down the contents of the tube; rinse off the paper or mortar with a small quantity of the hyposulphite mixture and pour into the tube; then fill up with soda-lime to within 2 inches of the end of the tube. (4) Place another plug of ignited asbestus at the end of the tube, and close with a cork. (5) Hold the tube in a horizontal position, and tap on the table until there is a gas channel all along the top of the tube. Make connection with the U tube containing the acid, aspirate, and see that the apparatus is tight.

The combustion.—Place the prepared combustion tube in the furnace, letting the open end project a little so as not to burn the cork. Commence by heating the soda-lime portion until it is brought to a full red heat. Then turn on slowly jet after jet toward the outer end of the tube, so that the bubbles come off two or three a second. When the whole tube is red hot and the evolution of the gas has ceased and the liquid in the U tube begins to recede towards the furnace, attach the aspirator to the other limb of the U tube, break off the end of the tube, and draw a current of air through for a few minutes. Detach the U tube and wash the contents into a beaker or porcelain basin, add a few drops of the cochineal solution, and titrate.

THE SODA-LIME METHOD.

(Not applicable in presence of nitrates.)

APPARATUS AND RE-AGENTS.

(1) Standard solutions and indicator the same as for the Kjeldahl method.

(2) Granulated soda-lime, fine enough to pass a 10-mesh seive, and thoroughly dry.

(3) Fine soda-lime, fine enough to pass a 20-mesh seive, also thoroughly dry.

To prepare soda-lime of the required fineness, the coarse granulated article of the trade may be ground until it will pass through a seive of about .10 inch mesh. It is then sifted on a seive of about .20-inch mesh to separate the fine from the coarse. The two portions are then thoroughly dried in the air-bath, or by heating in a porcelain dish or iron pan over a lamp, stirring constantly.

Excellent soda-lime may be easily and cheaply prepared according to the directions of Professor Atwater (Am. Chem. Journal, vol. 9, p. 312), by slaking $2\frac{1}{2}$ parts of quick-lime with a strong solution of 1 part commercial caustic soda (such soda as is used in the Kjeldahl process), care being taken that there is enough water in the solution to slake the lime. The mixture is then dried and heated in an iron pot to incipient fusion, and when cold, ground and sifted as above.

Instead of soda-lime, Johnson's mixture of carbonate of soda and lime or slaked lime may be used.

Slaked lime may be granulated by mixing it with a little water to form a thick mass, which is dried in the water-oven until hard and brittle. It is then ground and sifted as above. Slaked lime is much easier to work with than soda-lime and gives excellent results, though it is probable that more of it should be used, in proportion to the substance to be analyzed, than is the case with soda-lime.

(4) Asbestus which has been ignited and kept in a glass-stoppered bottle.

(5) Combustion tubes about 40 centimeters long and about 12 millimeters internal diameter, drawn out to a point and closed at one end.

(6) Large bulbed "U" tubes with glass stop-cock, or Wills tubes with four bulbs.

THE DETERMINATION.

The substance to be analyzed should be powdered fine enough to pass through a seive of 1 millimeter mesh; 0.7 to 1.4 grams, according to the amount of nitrogen present, is taken for the determination. Into the closed end of the combustion tube put a small, loose plug of asbestus, and upon it about 4 centimeters of fine soda-lime. In a porcelain dish or mortar mix the substance to be analyzed, thoroughly but quickly, with enough fine soda-lime to fill about 16 centimeters of the tube, or about 40 times as much soda-lime as substance, and put the mixture into the combustion tube as quickly as possible by means of a wide-necked funnel, rinsing out the dish and funnel with a little more fine soda-lime, which is to be put in on top of the mixture. Fill the rest of the tube to about 5 centimeters of the end with a granulated lime, making it as compact as possible by tapping the tube gently, while held in a nearly upright position during the filling. The layer of granulated soda-lime should be not less than 12 centimeters long. Lastly, put in a plug of asbestus about 2 centimeters long, pressed rather tightly, and wipe out the end of the tube to free it from adhering soda-lime.

Connect the tube by means of a well-fitting rubber stopper or cork with the "U" tube or Wills bulbs containing 10 cubic centimeters standard acid, and adjust it in the combustion furnace so that the end projects about 4 centimeters from the furnace, supporting the " U " tube or Wills tube suitably. Heat the portion of the tube containing the granulated soda lime to moderate redness, and when this is attained, extend the heat gradually through the portion containing the substance so as to keep up a moderate and regular flow of gases through the bulbs, maintaining the heat of the first part until the whole tube is heated uniformly to the same degree. Keep up the heat until gases have ceased bubbling through the acid in the bulbs and the mixture of substance and soda-lime has become white or nearly so, which shows that the combustion is finished. The combustion should occupy about three-quarters of an hour, or not more than one hour. Remove the heat, and when the tube has cooled below redness, break off the closed tip and aspirate air slowly through the apparatus for two or three minutes to bring all the ammonia into the acid. Disconnect, wash the acid into a beaker or flask, and titrate with the standard alkali.

During the combustion, the end of the tube projecting from the furnace must be kept heated sufficiently to prevent the condensation of moisture, yet not enough to char the stopper. The heat may be regulated by a shield of tin slipped over the projecting end of the combustion tube.

It is found very advantageous to attach a Bunsen valve to the exit tube, allowing the evolved gases to pass out freely, but preventing a violent "sucking back" in case of a sudden condensation of steam in the bulbs.

METHOD FOR THE ANALYSIS OF CATTLE FOOD.

Hygroscopic moisture.—Dry 2 to 3 grams at 100° C. to constant weight.

Ash.—Char the substance at a low red heat, exhaust the charred mass with water, burn the insoluble residue, add the ash to the residue from the evaporation of the aqueous extract obtained above, dry the whole at 110°, and weigh. Determine carbonic acid and insoluble matter (sand and charcoal) in the product for the estimation of the pure ash.

Ether extract.—Pulverize the air-dry substance till all of it will pass through a sieve with less than one-sixtieth inch mesh; dry about 5 grams at 100°, and take about 2 grams for each extraction. Exhaust with anhydrous ether of specific gravity .720 to .725, not less than eight hours continuously. Dry ether extract at 100° in a current of dry hydrogen to a constant weight.

Crude protein.—Determine nitrogen by the Kjeldahl method in the manner recommended by the nitrogen committee, and multiply the result by 6.25 for the crude protein.

Albuminoid nitrogen, Stutzer's method.—Prepare cupric hydrate as follows: Dissolve 100 grams of pure cupric sulphate in 5 liters of water and add 2.5 cubic centimeters of glycerine; add dilute solution of sodium hydrate till the liquid is alkaline, filter, rub the precipitate up with water containing 5 cubic centimeters of glycerine per liter, and then wash by decantation or filtration till the washings are no longer alkaline. Then rub the precipitate up again in a mortar with water containing 10 per cent. of glycerine, thus preparing a uniform gelatinous mass that can be measured out with a pipette. Determine the quantity of cupric hydrate per 1 cubic centimeter of this mixture.

To 1 gram of the substance, pulverized as for the determination of the ether extract, add 100 cubic centimeters of water in a beaker, heat to boiling, or in case of substances rich in starch heat on the water bath ten minutes, add a quantity of the cupric hydrate mixture containing 0.7 to 0.8 gram of the hydrate, stir thoroughly, filter when cold, wash with cold water, and without drying put the filter and its contents into the concentrated sulphuric acid for the determination of the nitrogen after Kjeldahl. For filtering use either Schleicher and Shull's No. 589 filter paper or Swedish paper, either of which contains so little nitrogen that it can be left out of account.

If the substance examined consists of seed of any kind or residues of seeds, such as oil-cake or anything rich in alkaline phosphate, add a few cubic centimeters of a concentrated solution of alum just before adding the cupric hydrate, and mix well by stirring. This serves to decompose the alkaline phosphate, precipitating aluminum phosphate; if this is not done cupric phosphate and pure alkali may be formed, and the protein copper may be partially dissolved in this alkaline liquid.

Crude fiber, Weende method.—Pulverize as for the ether extract, and extract the fat, at least nearly completely. To 2 grams of the substance in an Erlenmeyer flask add 200 cubic centimeters of boiling 1.25 per cent. sulphuric acid, boil immediately and for thirty minutes continuously with as much rotary motion as may be necessary to keep all the substances that may tend to adhere to the sides of the flask in contact with the liquid, throw on a filter of finest linen, and rinse the flask and wash the contents of the filter thoroughly with boiling water. Wash the contents of the filter back into the flask with 200 cubic centimeters of a 1.25 per cent. solution of sodium hydrate, at once raise to boiling, and boil continuously for thirty minutes with rotary motion as above described, filter through a Gooch crucible or according to some similar device, and wash very thoroughly with boiling water. Such thorough washing has been found to be essential to success. Wash then with alcohol and finally with ether, dry at 110°, weigh, incinerate, and give the loss of weight for crude fiber.

Statement of results.

	In the substance in its natural condition (or as received).	In the dry substance.
Moisture		
Crude ash		
Ether extract..................		
Crude fiber		
Crude protein		
Nitrogen: Total............		
Albuminoid		

METHODS OF ANALYSES OF DAIRY PRODUCTS.

BUTTER.

MICROSCOPIC EXAMINATION.

Place a small portion of the fresh sample, taken from the inside of the mass, on a slide, add a drop of pure sweet-oil, cover with gentle pressure, and examine with a one-half to one-eighth inch objective for crystals of lard, etc.

Examine same specimen with polarized light and selenite plate without the use of oil.

Pure fresh butter will neither show crystals nor a parti-colored field with selenite. Other fats, melted and cooled and mixed with butter, will usually present crystals and variegated colors with the selenite plate.

SPECIFIC GRAVITY.

Take a specific-gravity flask, not graduated, the stopper of which carries a delicate thermometer, whose bulb occupies sensibly the center of the flask. The thermometer should be first compared with a standard instrument.

Clean the picnometer with water, alcohol, and ether; dry, be careful to remove all the ether vapor, and weigh.

Fill with distilled water, insert stopper, and place in vessel containing water which comes nearly to top of picnometer (without stopper); gradually raise the temperature of water in vessel containing flask until it is 40.5° to 41° C.

The moment the thermometer of the picnometer marks 40° remove the flask, dry carefully, cover capillary safety-tube, and set in balance or desiccator to cool. Weigh when temperature falls nearly to that of balance room. Having determined the titre of the flask, it is to be carefully dried, filled with the melted and filtered fat at a temperature below 40°, and the weight of the fat determined at 40°, as directed above.

Apparatus.—The apparatus consists of (1) an accurate thermometer, for reading easily tenths of a degree; (2) a less accurate thermometer for measuring the temperature of water in the large beaker glass; (3) a tall beaker glass, 35 centimeters high and 10 centimeters in diameter; (4) a test tube 30 centimeters high and 3.5 centimeters in diameter; (5) a stand for supporting the apparatus; (6) some method of stirring the water in the beaker—for example, a blowing bulb of rubber and a bent glass tube extending to near the bottom of the beaker; (7) a mixture of alcohol and water of the same specific gravity as the fat to be examined.

Manipulation.—The disks of the fat are prepared as follows: The melted and filtered fat is allowed to fall from a dropping tube from a height of 15 to 20 centimeters on to a smooth piece of ice floating in water. The disks thus formed are from 1 to 1½ centimeters in diameter and weigh about 200 milligrams. By pressing the ice under the water the disks are made to float on the surface, whence they are easily removed with a steel spatula.

The mixture of alcohol and water is prepared by boiling distilled water and 95 per cent. alcohol for ten minutes to remove the gases which they may hold in solution. While still hot the water is poured into the test-tube already described until it is nearly half full. The test-tube is then filled with the hot alcohol. It should be poured in gently down the side of the inclined tube to avoid too much mixing. If the tube is not filled until the water has cooled the mixture will contain so many air bubbles as to be unfit for use. These bubbles will gather on the disk of fat as the temperature rises and finally force it to the top of the mixture.

The test-tube containing the alcohol and water is placed in a vessel containing cold water, and the whole cooled to below 10° C. The disk of fat is dropped into the tube from the spatula, and at once sinks until it reaches a part of the tube where the density of the alcohol-water is exactly equivalent to its own. Here it remains at rest and free from the action of any force save that inherent in its own molecules.

The delicate thermometer is placed in the test-tube and lowered until the bulb is just above the disk. In order to secure an even temperature in all parts of the alcohol mixture in the vicinity of the disk the thermometer is moved from time to time in a circularly pendulous manner. A tube prepared in this way will be suitable for use for several days; in fact, until the air bubbles begin to attach themselves to the disk of fat. In no case did the two liquids become so thoroughly mixed as to lose the property of holding the disk at a fixed point, even when they were kept for several weeks.

In practice, owing to the absorption of air, it has been found necessary to prepare new solutions every third or fourth day.

The disk having been placed in position, the water in the beaker-glass is slowly heated and kept constantly stirred by means of the blowing apparatus already described.

When the temperature of the alcohol-water mixture rises to about 6° below the melting-point, the disk of fat begins to shrivel, and gradually rolls up into an irregular mass.

The thermometer is now lowered until the fat particle is even with the center of the bulb. The bulb of the thermometer should be small, so as to indicate only the temperature of the mixture near the fat. A gentle rotary movement should be given to the thermometer bulb, which might be done with a kind of clockwork. The rise of temperature should be so regulated that the last 2° of increment require about ten minutes. The mass of fat gradually approaches the form of a sphere, and when it is sensibly so the reading of the thermometer is to be made. As soon as the temperature is taken the test-tube is removed from the bath and placed again in the cooler. A second tube, containing alcohol and water, is at once placed in the bath. It is not necessary to cool the water in the bath. The test-tube (ice water being used as a cooler) is of low enough temperature to cool the bath sufficiently. After the first determination, which should be only a trial, the temperature of the bath should be so regulated as to reach a maximum about 1°.5 above the melting-point of the fat under examination.

Working thus with two tubes about three determinations can be made in an hour.

After the test-tube has been cooled the globule of fat is removed with a small spoon attached to a wire before another disk of fat is put in.

VOLATILE ACIDS.

Take about 2.5 grams filtered fat in 200 cubic centimeters flask, to be used in subsequent distillation; add 2 cubic centimeters aqueous solution potassium hydrate containing .5 gram KOH per cubic centimeter. Add 2 cubic centimeters 95 per cent. alcohol and heat on water bath, with occasional agitation. Saponification is complete in a few minutes. Evaporate until alcohol is all driven off, using a current of air to remove alcoholic vapor. The flask is fitted with a delivery tube and condenser. The delivery tube is carried up about 8 inches before it is bent to enter the condenser, and a bulb is blown in it just below the elbow, and this is filled with broken glass or glass-wool.

The fatty acids are then set free with 20 cubic centimeters of a solution of phosphoric acid made by dissolving 200 grams glacial phosphoric acid in water and making up to 1 liter. Enough water is added to make the volume of liquid in the flask 70 cubic centimeters. Distillation is continued until the distillate measures 50 cubic centimeters, which is titrated with $\frac{N}{10}$ NaOH. using phenol-phthalein as indicator.

(Care should be taken to have the potash used free from carbonate.

A potash containing carbonate saponifies a fat with great difficulty.   In an instance where the saponification was taking place very slowly, the potash was found to contain 6.80 per cent. CO_2, equal to 21.17 per cent. K_2CO_3.)

SOLUBLE ACIDS.

The washings obtained in the process of estimating the insoluble acids, the description of which follows, are made up to 1 liter and an aliquot part taken for titration with one-tenth N alkali, using phenol-phthalein as indicator.

METHOD FOR THE DETERMINATION OF SOLUBLE AND INSOLUBLE ACIDS.

RE-AGENTS.

1. A standard semi-normal hydrochloric-acid solution, accurately prepared.

2. A standard decinormal soda solution, accurately prepared; each 1 cubic centimeter, contains .0040 gram of NaOH, and neutralizes .0088 gram of butyric acid, $C_4H_8O_2$.

3. An approximately semi-normal alcoholic potash. Dissolve 40 grams of good stick potash in 1 liter of 95 per cent. alcohol, redistilled. The solution must be clear and the KOH free from carbonates.

4. A 1 per cent solution of phenol-pthalein in 95 per cent. alcohol.

Saponification is carried out in rubber-stoppered beer-bottles holding about 250 cubic centimeters.

About 5 grams of the melted butter fat, filtered and freed from water and salt, are weighed out by means of a small pipette and beaker, which are reweighed after the sample has been taken out and run into a saponification bottle; 5) cubic centimeters of the semi-normal potash is added, the bottle closed and placed on the steam bath until the contents are entirely saponified, facilitating the operation by occasional agitation.   The alcoholic potash is measured always in the same pipette and uniformity further insured by always allowing it to drain the same length of time, viz, thirty seconds.   Two or three blanks are also measured out at the same time and treated in the same way.

In from five to thirty minutes, according to the nature of the fat, the liquid will appear perfectly homogeneous, and when this is the case the saponification is complete, and the bottle may be removed and cooled.   When sufficiently cool the stopper is removed and the contents of the flask rinsed with a little 95 per cent. alcohol into an Erlenmeyer flask of about 250 cubic centimeters capacity, which is placed on the steam bath, together with the blanks, until the alcohol has evaporated.

Titrate the blanks with semi-normal HCl, using phenol-phthalein as an indicator.   Then run into each of the flasks containing the fat acids 1 cubic centimeter more semi-normal HCl than is required to neutralize the potash in the blanks. The flask is then connected with a condensing tube 3 feet long, made of small glass tubing, and placed on the steam bath until the separated fatty acids form a clear stratum on the surface of the liquid. The flask and contents are then allowed to become thoroughly cold, ice water being used for cooling.

The fatty acids having quite solidified, the contents of the flask are filtered through a dry filter paper into a liter flask, care being taken not to break the cake. Two hundred to three hundred cubic centimeters of hot water is next poured on the contents of the flask, the cork with its condenser tube re-inserted, and heated on the steam bath until the cake of acids is thoroughly melted, the flask being occasionally agitated with a circular motion, so that none of its contents are brought on the cork. When the fatty acids have again separated as an oily layer the flask and its contents are cooled in ice water, and the liquid filtered through the same filter into the same liter flask. This treatment with hot water, followed by cooling and filtration of the wash water, is repeated three times, the washings being added to the first filtrate. The mixed washings and filtrate are next made up to 1 liter, and 100 cubic centimeters, in duplicate, are taken and titrated with decinormal NaOH. The volume required is calculated to the whole liquid. The number so obtained represents the measure of decinormal NaOH neutralized by the soluble fatty acids of the butter fat taken, plus that corresponding to the excess of the standard acid used, viz, 1 cubic centimeter. The amount of soda employed for the neutralization is to be diminished, for the 1 liter, by 5 cubic centimeters, corresponding to the excess of 1 cubic centimeter $\frac{1}{2}$ N. acid.

This corrected volume, multiplied by the factor 0.0088, gives the butyric acid in the weight of butter fat employed. (See table.)

The flask containing the cake of insoluble fat acids is inverted and allowed to drain and dry for twelve hours, together with the filter paper through which its soluble fatty acids have been filtered. When dry the cake is broken up and transferred to a weighed glass evaporating dish. Remove from the dried filter paper as much of the adhering fat acids as possible and add them to the contents of the dish. The funnel, with the filter paper, is then placed in an Erlenmeyer flask, a hole is made in the bottom of the filter paper, and it is thoroughly washed with absolute alcohol from a wash-bottle. The flask is rinsed with the washings from the filter paper and pure alcohol, and these transferred to the evaporating dish. The dish is placed on the steam bath and the alcohol driven off. It is then transferred to the air bath and dried at 100° C. for two hours, taken out, cooled in a desiccator, and weighed. It is then again placed in the air bath and dried for another two hours, cooled as before, and weighed. If there is no considerable decrease in weight the first weight will do; otherwise, reheat two hours and weigh. This gives the weight of insoluble fat acids in the quantity taken, from which the percentage is easily calculated.

Table for the calculation of soluble fatty acids.

No. cc. KOH sol.	Equiv.	$\frac{N}{10}$ NaOH.	Equiv.	$\frac{N}{10}$ NaOH.	Equiv.	$\frac{N}{10}$ NaOH.	Equiv.
	Grams.		Grams.		Grams.		Grams.
10	.0c8	+.25	.0902	+.50	.0921	+.75	.0946
11	.0968	25	.0990	50	.1012	75	.1034
12	.1056	25	.1078	50	.1100	75	.1122
13	.1144	25	.1166	50	.1188	75	.1210
14	.1232	25	.1254	50	.1276	75	.1298
15	.1320	25	.1342	50	.1364	75	.1386
16	.1408	25	.1430	50	.1452	75	.1474
17	.1496	25	.1518	50	.1540	75	.1562
18	.1584	25	.1606	50	.1628	75	.1650
19	.1672	25	.1694	50	.1716	75	.1738
20	.1760	25	.1782	50	.1804	75	.1826
21	.1848	25	.1870	50	.1892	75	.1914
22	.1936	25	.1958	50	.1980	75	.2002
23	.2024	25	.2046	50	.2068	75	.2090
24	.2112	25	.2134	50	.2156	75	.2178
25	.2200	25	.2222	50	.2244	75	.2266
26	.2288	25	.2310	50	.2332	75	.2354
27	.2376	25	.2398	50	.2420	75	.2442
28	.2464	25	.2486	50	.2508	75	.2530
29	.2552	25	.2574	50	.2590	75	.2618
30	.2640	25	.2662	50	.2684	75	.2706

The table gives the weight of soluble fatty acids (butyric, etc.) for each quarter of a cubic centimeter of decinormal alkali from 10 to 30.

Example: Weight fat taken .... grams.. 4.967

No. cubic centimeters $\frac{N}{10}$ alkali used 25.50

Less 5 cubic centimeters due to 1 cubic centimeter $\frac{N}{2}$ acid 20.50

Weight soluble fat acids .... .1504

Per cent. soluble fat acids 3.63

SAPONIFICATION EQUIVALENT.

About 2.5 grams butter fat (filtered and free from water) are weighed into a patent rubber-stoppered bottle, and 25 cubic centimeters approximately semi-normal alcoholic potash added. The exact amount taken is determined by weighing a small pipette with the beaker of fat, running the fat into the bottle from the pipette and weighing beaker and pipette again. The alcoholic potash is measured always in the same pipette, and uniformity further insured by always allowing it to drain the same length of time (thirty seconds). The bottle is then placed in the steam bath, together with a blank, containing no fat. After saponification is complete and the bottles cooled, the contents are titrated with accurately semi-normal hydrochloric acid, using phenol-phthalein as an indicator. The number of cubic centimeters of the acid used for the sample deducted from the number required for the blank gives the number of cubic centimeters which combines with the fat, and the saponification equivalent is calculated by the following formula, in which W equals the weight of fat taken in milligrams and N the number of cubic centimeters which have combined with the fat.

$$\text{Sap. Equiv.} = \frac{2\,W}{N}.$$

If it is desirable to express the number of milligrams of potash for each gram of fat employed it can be done by dividing 5,610 by the saponification equivalent and multiplying the quotient by ten.

WATER.

Place about 2 grams in flat-bottomed platinum dish two-thirds full of clean, dry sand and heat for two hours in air bath at 105° C. (For alternate method of estimation of water, see page 81.)

ESTIMATION OF SALT.

Volumetric method.—The amount of the butter or butter substitute to be taken is 5–10 grams; weigh in a counterpoised beaker-glass. The butter (fresh from the refrigerator) is placed in portions of about 1 gram at a time in the beaker, these portions being taken from different parts of the sample. By this means a reasonably fair sample of the whole is obtained.

The given quantity having been weighed out, it is removed from the pan.

Hot water is now added (about 20 cubic centimeters) to the beaker, containing the butter, and after it has melted the liquid is poured into the bulb of the separating apparatus. The stopper is now inserted and the contents shaken for a few moments.

After standing until the fat has all collected on top of the water, the stop-cock is opened and the water, containing most of the salt, is allowed to run into an Erlenmeyer flask, being careful to let none of the fat globules pass.

Hot water is again added to the beaker and thence poured into the separatory apparatus, the bottle well shaken, and the foregoing process is repeated ten to fifteen times, using each time 10 to 20 cubic centimeters of water.

The resulting washings contain all but a mere trace of the NaCl, originally present in the butter.

Estimation of NaCl in filtrate.—The chloride of sodium is now determined in the filtrate by a standard solution of $AgNO_3$, using a few drops of a saturated solution of potassium chromate as indicator.

TOTAL MINERAL MATTER. (ASH) AND CURD.

The methods of estimating curd depend on the principle of first drying a weighed portion of the butter, and afterwards extracting the fat with ether or petroleum. The residual mass is then weighed and the curd determined by loss on ignition. This process is carried on as follows:

Five to ten grams of butter are dried, at 100° C., for a few hours in a porcelain dish. The dried fat, &c., is filtered through a Gooch crucible, the contents of the dish all brought into the crucible and well washed with ether or light petroleum. The filter crucible is dried for two hours in air bath and weighed. The curd is then determined by loss of weight on ignition.

The total mineral matter is represented by the residue left after ig. nition.

Curd and insoluble substances may be estimated somewhat more accurately as follows :

Place the butter (5 to 10 grams) in a Gooch crucible; set the crucible on a pad of blotting-paper in an air bath (100° C.) for an hour, wash two or three times with ether or light petroleum. Dissolve the salt with hot water, dry at 100° C. and weigh. Any insoluble mineral matter and the ash of the curd is weighed with it in this method.

CASEIN.

Method of Babcock.—Ten grams of the dried butter are treated with light petroleum until all fat is removed. The residue is then ignited with soda-lime or treated by the Kjeldahl method.

METHODS FOR MILK ANALYSIS.

ESTIMATION OF WATER.

From a weighing bottle take 5 cubic centimeters milk, put in a weighed thin glass dish or dish lined with tinfoil one-third full of powdered asbestos; dry for two hours at 100° C. The temperature obtained in a boiling-water bath does not reach 100° C. The milk should be dried in an air bath, the temperature of which is carefully controlled.

ALTERNATE METHOD OF ESTIMATING WATER.

Evaporate 1 to 2 grams of milk in shallow watch glass or platinum dish on the water bath for thirty minutes. Dry for an hour at 100° C. and weigh.

ESTIMATION OF CASEIN.

Take 5 grams of milk, digest in Kjeldahl apparatus with 20 cubic centimeters H_2SO_4, and estimate ammonia in the usual way.

ALTERNATE METHOD OF ESTIMATION OF CASEIN.

Rub up in a mortar the thin disk containing the dried residue from the above process or remove the foil containing it and transfer to soda-lime combustion tube in the usual way.

The mortar and pestle must be well cleaned with the soda-lime and these cleanings placed in the tube.

Or the dish or tinfoil and its contents may be transferred to a digestion flask and the casein estimated by the method of Kjeldahl.

ESTIMATION OF FAT.

Method of Adams.—The kind of paper and the method of using it first proposed by Adams are as follows :

As for material, the only extra article is some stout white blotting-paper, known in the trade as "white demy blotting mill 428," weighing 38 pounds per ream. This should be in unfolded sheets, machine-cut into strips $2\frac{1}{4}$ inches wide and 22 inches long; each sheet in this manner cuts into seven strips.

I have tried other papers, but none have answered so well as this; it is very porous and just thick enough. Each of these strips is carefully rolled into a helical coil, for which purpose I use a little machine, made by myself, consisting of a stout double wire, cranked twice at right angles, and mounted in a simple frame. One end of the strip being thrust between the two wires, the handle is turned and the coil made with great facility. This may be done, for the nonce, on a glass rod, the size of a cedar pencil. Two points have to be carefully attended to; the paper must not be broken, and the coil must be somewhat loose, the finished diameter being a little under an inch. I am in the habit of rolling up a considerable number at a time and placing each within a brass ring as it is rolled, inscribing on one corner with a lead-pencil its own proper number.

These coils are next thoroughly dried, and I need hardly say the accuracy of the process depends upon this drying. This can be satisfactorily done in an ordinary air bath at 100° C., providing the bath be heated properly and the paper kept in it long enough. I found the common way of heating the thin bottom of the bath with a single jet not to answer. My bath is placed upon a stout iron surface, which is heated by a large ring of jets ; in this way the heat is evenly distributed over the whole of the bottom of the bath, and the papers, which are put in a cage frame of tinned-iron wire 5 by $2\frac{1}{2}$ inches and divided into eight partitions, get evenly and completely dried, if allowed to remain in the bath all night, and weighed in a weighing tube next morning, and their weights having been registered according to their numbers, stored away ready for use, as follows:

The milk to be examined is shaken, and with a pipette 5 cubic centimeters are discharged into a small beaker 2 inches high by $1\frac{1}{4}$ inches in diameter, of a capacity of about 30 cubic centimeters, weighing about 12 grams. This charged beaker is first weighed and then a paper coil gently thrust into the milk very nearly to the bottom. In a few minutes the paper sucks up nearly the whole of the milk. The paper is then carefully withdrawn by the dry extremity of the coil and gently reversed and stood, dry end downwards, on a clean sheet of glass. With a little dexterity all but the last fraction of a drop can be removed from the beaker and got on the paper. The beaker is again weighed and the milk taken got by difference. It is of importance to take up the whole of the milk from the beaker, as I am disposed to consider the paper has a selective action, removing the watery constituents of the milk by preference over the fat.

The charged paper is next placed in the water oven on the glass plate, milk-end upwards, and rough-dried. Mismanagement may possibly cause a drop to pass down through the coil onto the glass. This accident ought never to occur; but if it does it is revealed in a moment by inspection of the surface of the glass, and the experiment is thereby lost.

In about an hour it is rough-dried and in a suitable condition for the extraction of the fat.

The method of Adams has been thoroughly tried by the English chemists and has received the approval of the English Society of Public Analysts. It gives uniformly about .2 per cent. more fat in normal milk than the ordinary gravimetric methods.

The following modification of the process may be used:

The blotting-paper is replaced by thick filtering paper cut into strips 2 feet long and 2.5 inches wide. These are thoroughly extracted by ether or petroleum or alcohol.

One end of the strip of paper being held horizontally by a clamp or by an assistant, 5 cubic centimeters milk is run out by a pipette from a weighing bottle along the middle of the strip of filtering paper, being careful not to let the milk get too near the ends of the paper and to secure an even distribution of it over the whole length of the slip. The pipette is replaced in the weighing bottle and the whole reweighed, and thus the quantity of milk taken is accurately determined. The strip of paper is now hung up over a sand bath in an inclosed space high enough to receive it where the air has a temperature of 100° C. (*circa*). In two or three minutes the paper is thoroughly dry. It is at once, while still hot, rolled into a coil and placed, before cooling, in the extraction apparatus already described.

The fat is dissolved by ether or petroleum, collected in a weighed flask, and, after thorough drying, weighed.

The fat after extraction may also be estimated volumetrically, as described in the method of Morse.

From data which have been collected, it appears that the estimation of the fat in milk by the lactocrite is strictly comparable with the results of the Adams method. Those who have this instrument, therefore, can use it instead of the method given.

ALTERNATE METHOD OF ESTIMATING THE FAT IN MILK.

Method of Morse, Piggot, and Burton.—This method consists in the dehydration of the milk by means of anhydrous sulphate of copper; the extraction of the fat by means of the low-boiling products of petroleum; the saponification of the butter by means of an excess of a standard solution of potassium hydroxide in alcohol; and the determination of the excess of the alkali by means of a solution of hydrochloric acid. The following apparatus and re-agents are required:

(1) A porcelain mortar and pestle.

(2) An extraction tube, 14 or 15 millimeters in diameter, 220 millimeters in length, with funnel-shaped top. A straight chloride of calcium tube may be used for this.

(3) A 200 cubic centimeter Erlenmeyer flask, strong enough to be used with a filter pump.

(4) A suitable stand for holding the flask and extraction tube.

(5) Ten cubic centimeter pipettes.

(6) Weighing glasses with ground-glass stoppers.

(7) A low-boiling gasoline, distilling between 30° and 60° C.

(8) Dehydrated sulphate of copper.

(9) Semi-normal solution of potash in 95 per cent. alcohol.

(10) A semi-normal solution of hydrochloric acid.

Manipulation.—Place about 20 grams of the anhydrous copper sulphate, roughly measured in a copper spoon of the size to hold about that amount, in a porcelain mortar; make a cavity in the center of the mass with the pestle. Allow 10 cubic centimeters of the milk to run on to the copper sulphate, being careful that none of it touches the sides of the mortar. When the milk is nearly dry, grind the mass up with a little clean sand, transfer to the extraction tube, gently pressing it down in the tube by means of a glass rod. The lower portion of the extraction tube to be packed with clean cotton wool. The fat is extracted in the following way: 15 cubic centimeters of benzine are poured over the material in the extraction tube and drawn down, with the aid of the filter pump, until the whole of the mass to be extracted has become wet with the liquid, when the connection with the pump is closed; after about five minutes another portion of 15 cubic centimeters of benzine is poured into the tube and the whole of the liquid slowly drawn through with aid of the pump into the flask. Usually one extraction of this kind is sufficient to withdraw the whole of the butter, but for the sake of greater accuracy the process may be repeated two or three times.

Titration.—The benzine may be evaporated and the residual butter fat saponified with about 25 cubic centimeters of the approximately semi-normal potash. The residual alkali is determined by means of the semi-normal hydrochloric acid, using phenol-phthalein as indicator. The difference between the amount required in this process and the amount necessary to neutralize the quantity of alkali taken gives the amount of alkali required for the saponification. The number of milligrams of potash required for one gram of the fat is taken at 230. The fat may also be accurately titrated without evaporating the benzine.

ALTERNATE METHOD OF ESTIMATING WATER AND FAT IN MILK.

Method of Babcock.—In the bottom of a perforated test-tube is placed a clump of clean cotton; the tube is then filled three-quarters full of ignited asbestos, lightly packed, and a plug of cotton inserted over it. The tube and contents are weighed and the plug of cotton carefully removed and 5 grams of milk from a weighed pipette run into it, and the plug of cotton replaced. The tube connected at its lower end by a rubber tube and adapter with a filter pump is placed in a drying oven of a 100° C., and a slow current of dry air drawn through it until the water is completely expelled, which in no case requires more than two hours.

The tube containing the solids from the above operation is placed in an extraction apparatus and exhausted with ether in the usual way.

ALTERNATE METHOD OF ESTIMATING WATER AND FAT.

Method of Professor Macfarlane.—A glass tube 4 to 5 cubic centimeters in length and 2 centimeters in diameter, open at one end, drawn out to a tube 5 millimeters in diameter at the other end, is two-thirds filled with asbestos fiber, such as is used in manufacturing packing. It is dried in the water bath for several hours, cooled in the desiccator, and weighed. Ten cubic centimeters of the milk is then added from a pipette, which is completely absorbed by the asbestos. It is then weighed, the additional weight of the milk representing the amount taken. The tube, along with many others, is placed in a water bath with constant level and dried for ten or twelve hours (during the night) at a temperature of 90°. Next morning the tubes are cooled in the desiccator and weighed, the loss in weight being the moisture. The tubes are then placed in the Soxhlet extraction apparatus and exhausted with petroleum ether for

7717—No. 19——6

four hours. They are then removed and dried in a steam bath, cooled in desiccator, and weighed. The loss represents the butter fat.

THE ESTIMATION OF SUGAR.

The re-agents, apparatus, and manipulation necessary to give the most reliable results in milk sugar estimation are as follows:

Re-agents.—(1) *Basic plumbic acetate*, specific gravity 1.97. Boil a saturated solution of sugar of lead with an excess of litharge, and make it of the strength indicated above. One cubic centimeter of this will precipitate the albumens in 50 to 60 cubic centimeters of milk.

(2) *Acid mercuric nitrate.* Dissolve mercury in double its weight of nitric acid, specific gravity 1.42. Add to the solution an equal volume of water. One cubic centimeter of this re-agent is sufficient for the quantity of milk mentioned above. Larger quantities can be used without affecting the results of polarization.

(3) *Mercuric iodide with acetic acid.* KI 33.2 grams, $HgCl_2$ 13.5 grams, H, C_2H_3O 20 cubic centimeters, H_2O 64 cubic centimeters.

Apparatus.—(1) Pipettes marked at 59.5, 60, and 60.5 cubic centimeters. (2) Sugar flasks marked at 102.4 cubic centimeters. (3) Filters, observation tubes, and polariscope. (4) Specific gravity spindle and cylinder. (5) Thermometers.

Manipulation.—(1) The room and milk should be kept at a constant temperature. It is not important that the temperature should be any given degree. The work can be carried on equally well at 15° C., 20° C., or 25° C. The slight variations in rotary power within the above limits will not affect the result for analytical purposes. The temperature selected should be the one which is most easily kept constant.

(2) The specific gravity of the milk is determined. For general work this is done by a delicate specific-gravity spindle. Where greater accuracy is required, use specific-gravity flask.

(3) If the specific gravity be 1.026, or nearly so, measure out 60.5 cubic centimeters into the sugar-flask. Add 1 cubic centimeter of mercuric nitrate solution, or 30 cubic centimeters mercuric iodide solution, and fill to 102.4 cubic-centimeter mark. The precipitated albumen occupies a volume of about 2.44 cubic centimeters. Hence the milk solution is really 100 cubic centimeters. If the specific gravity is 1.030, use 60 cubic centimeters of milk. If specific gravity is 1.034, use 59.5 cubic centimeters of milk.

(4) Fill up to mark in 102.4 cubic-centimeter flask, shake well, filter, and polarize.

NOTES.—In the above method of analysis the specific rotatory power of milk sugar is taken at 52.5, and the weight of it in 100 cubic centimeter solution to read 100 degrees in the cane-sugar scale at 20.56 grams. This is for instruments requiring 16.19 grams sucrose to produce a rotation of 100 sugar degrees. It will be easy to calculate the number for milk-sugar, whatever instrument is employed.

Since the quality of milk taken is three times 20.56 grams, the polariscopic readings divided by 3 give at once the percentage of milk sugar when a 200-millimeter tube is used.

If a 400-millimeter tube is employed, divide reading by 6; if a 500-millimeter tube is used, divide by 7.5.

Since it requires but little more time, it is advisable to make the analysis in duplicate and take four readings for each tube. By following this method gross errors of observation are detected and avoided.

By using a flask graduated at 102.4 for 60 cubic centimeters no correction for volume of precipitated caseine need be made. In no case is it necessary to heat the sample before polarizing.

In the above method no account is taken of the fat which is retained on the filter with the caseine. It is worth while to inquire if a correction similar to that made for the albuminoids should not also be made for the fat? (See page 10, Vieth's Investigation of this subject from Analyst 13, 63.)

ESTIMATION OF ASH.

Evaporate to dryness in a weighed platinum dish 20 cubic centimeters of milk from a weighing-bottle, to which 6 cubic centimeters of HNO_3 has been added, and burn in muffle at low red heat until ash is free from carbon.

METHODS FOR THE ANALYSIS OF FERMENTED LIQUORS.

A commission of experts, appointed in the year 1884 by the chancellor of the Empire, to which was intrusted the establishment of uniform methods for the chemical investigation of wine, adopted the following resolutions, which were made public by the Prussian minister for commerce and trade by a decree of the 12th August, 1884, which provides that they shall be rigidly adhered to in public institutions for the examination of food-stuffs, and are recommended to the representatives of like private concerns:

RESOLUTIONS OF THE COMMISSION FOR ESTABLISHING UNIFORM METHODS FOR THE ANALYSIS OF WINES. [1]

Since, in consequence of improper manner of taking, keeping, and sending in of samples of wine for investigation by the authorities, a decomposition or change in the latter often occurs, the commission considers it advisable to give the following instructions:

INSTRUCTIONS FOR SAMPLING, PRESERVING, AND SENDING IN OF SAMPLES OF WINE FOR EXAMINATION BY THE AUTHORITIES.

(1) Of each sample, at least one bottle (¾ liter), as well filled as possible, must be taken.

(2) The bottles and corks used must be perfectly clean; the best are new bottles and corks. Pitchers or opaque bottles in which the presence of impurities can not be seen are not to be used.

---

[1] Das Gesetz betreffend den Verkehr mit Nahrungsmittel, u. s. w., p. 184.

(3) Each bottle shall be provided with a label, gummed (not tied) on, upon which shall be given the index number of the sample corresponding to a description of it.

(4) The samples are to be sent to the chemical laboratory as soon as possible to avoid any chance of alteration which, under some circumstances, can take place in a very short time. If they are, for some special reason retained in any other place for any length of time, the bottles are to be placed in a cellar and kept lying on their sides.

(5) If in samples of wine taken from any business concern adulteration is shown, a bottle of the water is to be taken which was presumably used in the adulteration.

(6) It is advisable, in many cases necessary, that, together with the wine, a copy of these resolutions be sent to the chemist.

A.—*Analytical methods.*

Specific gravity.—In this determination use is to be made of a picnometer, or a Westphal balance controlled by a picnometer. Temperature, 15° C.

Alcohol.—The alcohol is estimated in 50 to 100 cubic centimeters of the wine by the distillation method. The amount of alcohol is to be given in the following way : In 100 cubic centimeters wine at 15° C. are contained *n* grams alcohol. For the calculation the tables of Baumhauer or Hebuer are used.

(The amounts of all the other constituents are also to be given in this way ; in 100 cubic centimeters wine at 15° C. are contained *n* grams.)

Extract.—For this estimation 50 cubic centimeters of wine, measured out at 15° C., are evaporated on the water bath in a platinum dish (85 millimeters in diameter, 20 millimeters in height, and 75 cubic centimeters capacity, weight about 20 grams), and the residue heated for two and one-half hours in a water jacket. Of wines rich in sugar (that is, wines containing over 0.5 grams of sugar in 100 cubic centimeters) a smaller quantity, with corresponding dilution, is taken, so that 1 or at the most 1.5 grams extract are weighed.

Glycerine.—One hundred cubic centimeters of wine (for sweet wines see below) are evaporated in a roomy, not too shallow, porcelain dish to about 10 cubic centimeters, a little sand added, and milk of lime to a strong alkaline reaction, and the whole brought nearly to dryness. The residue is extracted with 50 cubic centimeters of 96 per cent. alcohol on the water bath, with frequent stirring. The solution is poured off through a filter, and the residue exhausted by treatment with small quantities of alcohol. For this 50 to 100 cubic centimeters are generally sufficient, so that the entire filtrate measures 100 to 200 cubic centimeters. The alcoholic solution is evaporated on the water bath to a sirupy consistence. (The principal part of the alcohol may be distilled off if desired.) The residue is taken up by 10 cubic centimeters of absolute alcohol, mixed in a stoppered flask with 15 cubic centimeters of ether and allowed to stand until clear, when the clear liquid is poured off into a glass-stoppered weighing-glass, filtering the last portions of the solution. The solution is then evaporated in the weighing-glass until the residue no longer flows readily, after which it is dried an hour longer in a water jacket. After cooling it is weighed.

In the case of sweet wines (over 0.5 grams sugar in 100 cubic centimeters) 50 cubic centimeters are taken in a good-sized flask, some sand added, and a sufficient quantity of powdered slack lime, and heated with frequent shaking in the water bath. After cooling, 100 cubic centimeters of 96 per cent. alcohol are added, the precipitate which forms allowed to separate, the solution filtered, and the residue washed with alcohol of the same strength. The alcoholic solution is evaporated and the residue treated as above.

Free acids (total quantity of the acid reacting constituents of the wine).—These are to be estimated with a sufficiently dilute normal solution of alkali (at least one-third normal alkali) in 10 to 20 cubic centimeters wine. If one-tenth normal alkali is used at least 10 cubic centimeters of wine should be taken for titration; if one-third normal, 20 cubic centimeters of wine. The drop method (*Tupfel methode*), with delicate re-agent paper, is recommended for the establishment of the neutral point. Any considerable quantities of carbonic acid in the wine are to be previously removed by shaking. These "free acids" are to be reckoned and reported as tartaric acid ($C_4H_6O_6$).

Volatile acids.—These are to be estimated by distillation in a current of steam, and not indirectly, and reported as acetic acid ($C_2H_4O_2$). The amount of the "fixed acids" is found by subtracting from the amount of "free acids" found, the amount of tartaric acid corresponding to the "volatile acids" found.

Bitartrate of potash and free tartaric acid.—(*a*) Qualitative detection of free tartaric acid: 20 to 30 cubic centimeters of the wine are treated with precipitated and finely-powdered bitartrate of potash, shaken repeatedly, filtered off after an hour, and 2 to 3 drops of a 20 per cent. solution of acetate of potash added to the clear filtrate, and the solution allowed to stand twelve hours. The shaking and standing of the solution must take place at as nearly as possible the same temperature. If any considerable precipitate forms during this time free tartaric acid is present, and the estimation of it and of the bitartrate of potash may be necessary.

(*b*) Quantitative estimation of the bitartrate of potash and free tartaric acid: In two stoppered flasks two samples of 20 cubic centimeters of wine each are treated with 200 cubic centimeters ether-alcohol (equal volumes), after adding to the one flask 2 to 3 drops of a 20 per cent. solution of acetate of potash. The mixtures are well shaken, and allowed to stand sixteen to eighteen hours at a low temperature (0 — 10° C.), the precipitate filtered off, washed with ether-alcohol, and titrated. (The solution of acetate of potash must be neutral or acid. The addition of too much acetate may cause the retention of some bitartrate in solution.) It is best on the score of safety to add to the filtrate from the estimation of the total tartaric acid a further portion of 2 drops of acetate of potash, to see if a further precipitation takes place.

In special cases the following procedure of Nessler and Barth may be used as a control:

Fifty cubic centimeters of wine are evaporated to the consistency of a thin sirup (best with the addition of quartz sand), the residue brought into a flask by means of small washings of 96 per cent. alcohol, and with continual shaking more alcohol is gradually added, until the entire quantity of alcohol is about 100 cubic centimeters. The flask and contents are corked and allowed to stand four hours in a cool place, then filtered, and the precipitate washed with 96 per cent. alcohol; the filter paper, together with the partly flocculent, partly crystalline precipitate, is returned to the flask, treated with 30 cubic centimeters warm water, titrated after cooling, and the acidity reckoned as bitartrate. The result is sometimes too high if pectinous bodies separate out in small lumps, inclosing a small portion of free acids.

In the alcoholic filtrate the alcohol is evaporated, 0.5 cubic centimeters of a 20 per cent. potassic acetate solution added, which has been acidified by a slight excess of acetic acid, and thus the formation of bitartrate from the free tartaric acid in the wine facilitated. The whole is now, like the first residue of evaporation, treated with (sand and) 96 per cent. alcohol, and carefully brought into a flask, the volume of alcohol increased to 100 cubic centimeters, well shaken, corked, allowed to stand in a cold place four hours, filtered, the precipitate washed, dissolved in warm water, titrated, and for one equivalent of alkili two equivalents of tartaric acid are reckoned.

This method for the estimation of the free tartaric acid has the advantage over the former of being free from all errors of estimation by difference. The presence of considerable quantities of sulphates impairs the accuracy of the method.

Malic acid, succinic acid, citric acid.—Methods for the separation and estimation of these acids can not be recommenced at the present time.

Salicylic acid.—For the detection of this, 100 cubic centimeters of wine are repeatedly shaken out with chloroform, the latter evaporated and the aqueous solution of the residue tested with very dilute solution of ferric chloride. For the approximately quantitative determination it is sufficient to weigh the chloroform residue, after it has been again recrystalized from chloroform.

Coloring matter.—Red wines are always to be tested for coal-tar colors. Conclusions in regard to the presence of other foreign coloring matters drawn from the color of precipitates and other color reactions are only exceptionally to be regarded as safe. In the search for coal-tar colors the shaking out of 100 cubic centimeters of the wine with ether before and after its neutralization with ammonia is recommended. The etherial solutions are to be tested separately.

Tannin.—In case a quantitative determination of tannin (or tannin and coloring matter) appears necessary, the permanganate method of Neubauer is to be employed. As a rule the following estimation of the amount of tannin will suffice : The free acids are neutralized to within 0.5 grams in 100 cubic centimeters with standard alkali, if necessary. Then 1 cubic centimeter of 40 per cent. sodic acetate solution is added, and drop by drop a 10 per cent. solution of ferric chloride, avoiding an excess. One drop of ferric chloride is sufficient for the precipitation of 0.05 per cent. of tannin. (New wines are deprived of the carbonic acid held in solution by repeated shaking.)

Sugar.—The sugar should be determined, after the addition of carbonate of soda, by means of Fehling's solution, using dilute solutions, and, in wines rich in sugar (*i. e.*, wines containing over .5 gram in 100 cubic centimeters), with observance of Soxhlet's modifications, and calculated as grape sugar. Highly colored wines are to be decolorized with animal charcoal if their content of sugar is low, and with acetate of lead and sodium carbonate if it is high.

If the polarization indicates the presence of cane sugar (compare under polarization) the estimation is to be repeated in the manner indicated after the inversion (heating with hydrochloric acid) of the solution. From the difference the cane sugar can be calculated.

Polarization.—(1) With white wines: 60 cubic centimeters of wine are treated with 3 cubic centimeters acetate of lead solution in a graduated cylinder, and the precipitate filtered off. To 30 cubic centimeters of the filtrate is added 1.5 cubic centimeters of a saturated solution of sodic carbonate, filtered again, and the filtrate polarized. This gives a dilution of 10 : 11 which must be allowed for.

(2) With red wines: 60 cubic centimeters wines are treated with 6 cubic centimeters acetate of lead, and to 30 cubic centimeters of the filtrate 3 cubic centimeters of the saturated solution of sodic carbonate added, filtered again, and polarized. In this way a dilution of 5 : 6 is obtained.

The above conditions are so arranged (with white and red wines) that the last filtrate suffices to fill the ·220-millimeter tube of the Wild polaristrobometer of which the capacity is about 28 cubic centimeters.

In place of the acetate of lead very small quantities of animal charcoal can be used. In this case an addition of sodic carbonate is not necessary, nor is the volume of the wine altered. If a portion of the undiluted wine 220 millimeters long shows a higher right-handed rotation than 0.3°, Wild, the following procedure is necessary :

Two hundred and ten cubic centimeters of the wine are evaporated on the water bath to a thin sirup, after the addition of a few drops of a 20 per cent. solution of acetate of potash. To the residue is added gradually, with continual stirring, 200 cubic centimeters of 90 per cent. alcohol. The alcoholic solution, when perfectly clear, is poured off or filtered into a flask, and the alcohol distilled or evaporated off down to about 5 cubic centimeters. The residue is treated with about 15 cubic centimeters water and a little bone-black, filtered into a graduated cylinder, and washed with water until the filtrate measures 30 cubic centimeters. If this shows on polarization a rotation of more than +0.5°, Wild, the wine contains the unfermentable matter of commercial potato sugar (amylin). If in the estimation of the sugar by Fehling's solution more than 0.3 grams sugar in 100 cubic centimeters was found, the original right-rotation caused by the amylin may be diminished by the left-rotating sugar; the above precipitation with alcohol is in this case to be undertaken, even when the right-rotation is less than 0.3°, Wild. The sugar is, however, first fermented by the addition of pure yeast. With very considerable content in (Fehling's solution) reducing sugar and proportionally small left-rotation, the diminishing of the left-rotation may be brought about by cane sugar or dextrin or amylin. For the detection of the first the wine is inverted by heating with hydrochloric acid (to 50 cubic centimeters wine, 5 cubic centimeters dilute hydrochloric acid of specific gravity 1.10), and again polarized. If the left-rotation has increased, the presence of cane sugar is demonstrated. The presence of dextrin is shown as given in the section on "gum." In case cane sugar is present well washed yeast, as pure as possible, should be added, and the wine polarized after fermentation is complete. The conclusions are then the same as with the wines poor in sugar.

For polarization only large, exact instruments are to be used.

The rotation is to be calculated in degrees, Wild, according to Landolt (Zeitschr. f. analyt. Chemie, 7. 9):

1° Wild   = 4.6043° Soleil.
1° Soleil = 0.217189° Wild.
1° Wild   = 2.89005° Ventzke.
1° Ventzke = 0.346015° Wild.

Gum (arabic).—For establishing the addition of any considerable quantities of gum 4 cubic centimeters wine are treated with 10 cubic centimeters of 96 per cent. alcohol. If gum is present, the mixture becomes milky, and only clears up again after several hours. The precipitate which occurs adheres partly to the sides of the tube, and forms hard lumps. In genuine wine, flakes appear after a short time, which soon settle, and remain somewhat loose. For a more exact test it is recommended to evaporate the wine to the consistency of a sirup, extract with alcohol, of the strength given above, and dissolve the insoluble residue in water. This solution is treated with some hydrochloric acid (of specific gravity 1.10) heated under pressure two hours, and the reducing power ascertained with Fehling's solution, and calculated to dextrose. In genuine wines no considerable reduction is obtained in this way. (Dextrin is to be detected in the same way.)

Mannite.—As the presence of mannite in wines has been observed in a few cases, it should be considered when pointed crystals make their appearance in the extract or the glycerine.

Nitrogen.—In the estimation of nitrogen the soda-lime method is to be used.

Mineral matters.—For their estimation 50 cubic centimeters of wine are used. If the incineration is incomplete, the charcoal is leached with some water, and burned by itself. The solution is evaporated in the same dish, and the entire ash gently ignited.

Chlorine estimation.—The wine is saturated with sodic carbonate, evaporated, the residue gently ignited and exhausted with water. In this solution the chlorine is to be estimated volumetrically according to Volhard, or gravimetrically. Wines whose ashes do not burn white by gentle ignition usually contain considerable quantities of chlorine (salt).

Sulphuric acid.—This is to be estimated directly in the wine by the addition of barium chloride. The quantitative estimation of the sulphuric acid is to be carried out only in cases where the qualitative test indicates the presence of abnormally large quantities. (In the case of viscous or very muddy wines a previous clarification with Spanish earth is to be recommended.)

If in a special case it is necessary to investigate whether free sulphuric acid or potassium bisulphate are present, it must be proved that more sulphuric acid is present than is necessary to form neutral salts with all the bases.

Phosphoric acid.—In the case of wines whose ashes do not react strongly alkaline the estimation is made by evaporating the wine with sodic carbonate and potassic nitrate, the residue gently ignited and taken up with dilute nitric acid; then the molybdenum method is to be used. If the ash reacts strongly alkaline the nitric-acid solution of it can be used directly for the phosphoric-acid determination.

88

The other mineral constituents of wine (also alum) are to be determined in the ash or residue of incineration.

Sulphurous acid.—One hundred cubic centimeters wine are distilled in a current of carbonic acid gas after the addition of phosphoric acid. For receiving the distillate 5 cubic centimeters of normal iodine solution are used. After the first third has distilled off, the distillate, which must still contain an excess of free iodine, is acidified with hydrochloric acid, heated and treated with barium chloride.

Adulteration of grape wine with fruit wine.—The detection of this adulteration can only exceptionally be carried out with certainty by means of the methods that have so far been offered. Especially are all methods untrustworthy which rely upon a single reaction to distinguish grape from fruit wine; neither is it always possible to decide with certainty from the absence of tartaric acid, or from the presence of only very small quantities, that a wine is not made from grapes.

In the manufacture of artificial wine together with water the following articles are known to be sometimes used: Alcohol (direct or in the shape of fortified wine), cane sugar, starch sugar, and substances rich in sugar (honey), glycerine, bitartrate of potash, tartaric acid, other vegetable acids, and substances rich in such acids, salicylic acid, mineral matters, gum arabic, tannic acid, and substances rich in the same (e. g., kino, *catechu*), foreign coloring matters, various ethers and aromas.

The estimation, or rather the means of detecting the most of these substances has already been given above, with the exception of the aromas and others, for which no method can as yet be recommended.

The following substances may be mentioned here in particular, which serve for increasing the sugar, extract and free acid: Dried fruit, tamarinds, St. John's bread, dates, figs.

B.—*Rules for judging of the purity of wine.*

I. (a) Tests and determinations which are, as a rule, to be performed in judging of the purity of wines: Extract, alcohol, sugar, free acids as a whole, free tartaric acid qualitative, sulphuric acid, total ash, polarization, gum, foreign coloring matters in red wines. (b) Tests and determinations which are also to be carried out under special circumstances: Specific gravity, volatile acids, bitartrate of potash, and free tartaric acid quantitative, succinic acid, malic acid, citric acid, salicylic acid, sulphurous acid, tannin, mannite, special ash constituents, nitrogen.

The commission considers it desirable, in giving the estimations generally performed, to adhere to the order of succession given above, (under (a)).

II. The commission can not regard it as their province to give a guide for judging of the purity of wine, but thinks it advisable, in the light of its experience, to call attention to the following points:

Wines which are made wholly from pure grape juice very seldom contain a less quantity of extract than 1.5 grams in 100 cubic centimeters wine. If wines poorer in extract occur they should be condemned, unless it can be proven that natural wines of the same district and vintage occur with a similar low content of extract.

After subtracting the "fixed acids" the remaining extract (*extractrest*) in pure wines, according to previous experience, amounts to at least 1.1 grams in 100 cubic centimeters, and after substracting the "free acids," at least 1 gram. Wines which show less *extractrest* are to be condemned, in case it can not be shown that natural wines of the same district and vintage contain as small an *extractrest*.

A wine which contains appreciably more ash than 10 per cent. of its extract content must contain, correspondingly, more extract than would otherwise be accepted as a minimum limit. In natural wines the relation of ash to extract approaches very closely 1 to 10 parts by weight. Still a considerable deviation from this relation does not entirely justify the conclusion that the wine is adulterated.

The amount of free tartaric acid in pure wines, according to previous experience, does not exceed one-sixth of the entire "fixed acids."

The relation between alcohol and glycerine can vary in pure wines between 100 parts by weight of alcohol to 7 parts by weight of glycerine, and 100 parts by weight of alcohol to 14 parts by weight of glycerine. In case of wines showing a different glycerine relation an addition of alcohol or glycerine can be inferred.

As sometimes during its handling in cellars small quantities of alcohol (at most 1 per cent. by volume) may find their way into wine, this fact must be borne in mind in judging of its purity.

These proportions are not always applicable to sweet wines.

For the individual ash constituents no generally applicable limits can be given. The opinion that the better kinds of wine always contain more phosphoric acid than others is unfounded.

Wines that contain less than 0.14 gram of mineral matter in 100 cubic centimeters are to be condemned, if it can not be shown that natural wines of the same kind and the same vintage, which have been subject to like treatment, have an equally small content of mineral matter.

Wines which contain more than 0.05 gram of salt in 100 cubic centimeters are to be condemned.

Wines that contain more than 0.092 gram sulphuric acid (SO_3) corresponding to 0.20 grams potassic sulphate (K_2SO_4) in 100 cubic centimeters, are to be designated as wines containing too much sulphuric acid, either from the use of gypsum or in some other way.

Through various causes wines may become viscous, black, brown, cloudy, or bitter; they may otherwise change essentially in color, taste, and odor. The color of red wines may also separate in a solid form; still all these phenomena in and of themselves would not justify the condemnation of the wine as not genuine.

If during the summer time an energetic fermentation commences in a wine, this does not justify the conclusion that an addition of sugar or substances rich in sugar, e. g., honey, etc., has taken place, for the first fermentation may have been hindered in various ways or the wine may have had an addition of a wine rich in sugar.

The methods adopted by the "Union of Bavarian Chemists" differ considerably from the above in many particulars, so they are given also, together with the methods adopted by the same body for the examination of beer[1] in somewhat condensed form.

WINES.

METHODS OF INVESTIGATION.

I. *Determination of specific gravity.*—This is to be done by means of a Westphal's balance or a picnometer, and always at 15° C.

II. *Determination of extract.*—Ten to 50 cubic centimeters wine at 15° C. are evaporated in a platinum dish on the water bath to the proper consistence, and then dried in a drying oven at 100° C. to constant loss of weight. Constant loss of weight is assumed when three weighings, with equal intervals between the first and second and second and third give equal differences between the successive weighings.

Weighings are to be made at intervals of fifteen minutes.

III. *Inorganic matter.*—This is the incombustible ash obtained by burning the extract. Repeated moistening, drying, and heating to redness are advisable to entirely get rid of all organic constituents.

IV. *Acidity.*—After shaking vigorously, to drive off carbonic acid, the wine is to be titrated with an alkali solution and the acidity expressed in terms of tartaric acid.

V. *Glycerine.*—(1) This is determined in dry wines as follows : The alcohol is driven off from 100 cubic centimeters wine, lime or magnesia added, and the mass evaporated to dryness. The residue is boiled with 90 per cent. alcohol, filtered, and the

[1] Hilger, Vereinbarnngen u. s. w., p. 154.

filtrate evaporated to dryness. This residue is dissolved in 10 to 20 cubic centimeters alcohol, 15 to 30 cubic centimeters ether added, and the mixture allowed to stand until it is clear. It is then decanted from the sticky precipitate into a glass-stoppered weighing-bottle, evaporated to constant loss of weight, and weighed.

(2) The following method is employed for sweet wines : One hundred cubic centimeters wine are measured into a porcelain dish and evaporated on the water bath to a sirupy consistence, mixed with 100 to 150 cubic centimeters absolute alcohol, poured into a flask, ether added in the proportion of 1½ volumes to each volume of alcohol used, the flask well shaken, and allowed to stand until the liquid becomes clear. This is then poured off and the residue again treated with a mixture of alcohol and ether. The liquids are mixed, the alcohol and ether driven off, the residue dissolved in water, and treated as in (1).

(3) In all glycerine determinations it is necessary to take into consideration the loss of glycerine due to its volatility with water and alcohol vapor, and accordingly to add to the glycerine found 0.100 gram for each 100 cubic centimeters of liquid evaporated.

(4) It is necessary to test the glycerine from sweet wines for sugar, and if any is present it must be estimated by Soxhlet's or Knapp's method and its weight subtracted from that of the glycerine.

VI. *Alcohol.*—The determination must be made by distillation in glass vessels, and the results stated as follows : One hundred cubic centimeters wine at 15° C. contain x grams or cubic centimeters alcohol.

VII. *Polarization.*—(1) The wine is decolorized with plumbic subacetate.

(2) A slight excess of sodic carbonate is added to the filtrate from (1). Two cubic centimeters of a solution of plumbic subacetate are added to 40 cubic centimeters white wine and 5 cubic centimeters to 40 cubic centimeters red wine, the solution is filtered and 1 cubic centimeter of a saturated solution of sodic carbonate added to 21 or 22.5 cubic centimeters of the filtrate.

(3) The kind of apparatus used and the length of the tube are to be given, and results estimated in equivalents of Wild's polaristrobometer with 200-millimeter tubes.

(4) All samples rotating more than 0.5 degrees to the right (in 220-millimeter tubes, after treating as above), and showing no change, or but little change, in their rotatory power after inversion, are to be considered as containing unfermented glucose (starch sugar) residue.

(5) Rotatory power of less than 0.3 degrees to the right shows that impure glucose has not been added.

(6) Wines rotating between 0.3 degrees and 0.5 degrees to the right must be treated by the alcohol method.

(7) Wines rotating strongly to the left must be fermented and their optical properties then examined.

VIII. *Sugar.*—This is to be determined by Soxhlet's or Knapp's method. The presence of unfermented cane sugar is to be shown by inversion, etc.

IX. *Potassic bitartrate.*—The determination of potassic bitartrate as such is to be omitted.

X. *Tartaric, malic, and succinic acids.*—(1) According to Schmidt and Hiepe's method.

(2) Determination of tartaric acid according to the modified Berthelot-Fleury method.

(3) If the addition of 1 gram finely-powdered tartaric acid to 100 grams wine produces no precipitate of potassic bitartrate, the modified Berthelot-Fleury method must be employed to determine free tartaric acid.

XI. *Coloring matter.*—(1) Only aniline dyes are to be looked for.

(2) Special attention is to be paid to the spectroscopic behavior of rosaniline dyes, as obtained by shaking wines with amyl alcohol before and after saturation with ammonia.

(3) A qualitative test for alumina is not sufficient evidence of the addition of alum.

XII. *Nitrogen.*—To be determined according to the ordinary method.

XIII. *Citric acid.*—Presence to be shown by a qualitative test, as baric citrate.

XIV. *Sulphuric acid.*—To be determined in the wine after adding hydrochloric acid.

XV. *Chlorine.*—To be determined in the nitric-acid solution of the burnt residue by Volhard's method.

XVI. *Lime, magnesia, and phosphoric acid.*—These are determined in the ash fused with sodic hydrate and potassic nitrate, the phosphoric acid by the molybdenum method.

XVII. *Potash.* Either in the wine ash, as the platinum double salt, or in the wine itself, by Kayser's method.

XVIII. *Gums.*—Presence shown by precipitation by alcohol; 4 cubic centimeters wine and 10 cubic centimeters 96 per cent. alcohol are mixed. If gum arabic has been added, a lumpy, thick, stringy precipitate is produced; whereas pure wine becomes at first opalescent and then flocculent.

METHODS OF JUDGING PURITY—(Beurtheilung).

Part I.

I. Commercial wines may be defined as follows: (*a*) The product obtained by the fermentation of grape juice with or without grape skins and stems. (*b*) The product obtained by the fermentation of pure must, to which pure sugar, water, or infusion of grape skins has been added. It must contain not more than 9 per cent. alcohol and 0.3 per cent. sugar, and not less than 0.7 per cent. acid, estimated as tartaric. (*c*) The product obtained in southern countries by the addition of alcohol to fermented or partly fermented grape juice. French wines are not included, however. (*d*) The product obtained by fermenting the expressed juice of more or less completely dried wine grapes.

II. The above definitions do not apply to champagnes.

III. The following include the operations undergone by wines in cellars (*Kellermässige Behandlung*): (*a*) Drawing and filling. (*b*) Filtration. (*c*) Clarification by the use of kaolin, isinglass, gelatine or albumin, with or without tannin. (*d*) Sulphuring. Only minute traces of sulphurous acid may be contained in wine for consumption. (*e*) Adulteration of wine. (*f*) Addition of alcohol to wine intended for export.

IV. Wines, even if plastered, must not contain more sulphuric acid than that corresponding to 2 grams potassic sulphate (K_2SO_4) per liter.

V. Medicinal wines are those mentioned in Parts I and IV, with the following restrictions: (*a*) They must not contain more sulphuric acid than corresponds to 1 gram potassic sulphate per liter. (*b*) They must contain no sulphurous acid. (*c*) The percentage of alcohol and sugar to be given on the label. (*d*) These restrictions apply only to wines expressly recommended or sold for medicinal use.

Part II.

I. Improperly gallized wines are preparations of grape juice, pure sugar and water, or grape-skin infusion, that contain more than 9 per cent. alcohol or less than 0.7 per cent. acid, or both, and preparations in which impure glucose has been used. The following facts enable us to detect them: Small quantity of inorganic matter (phosphoric acid and magnesia), and right rotation if impure glucose is used. If the rotation exceeds 0.2° to the right, the wine is to be concentrated, freed from tartaric acid as far as possible, and again polarized.

II. Addition of alcohol is to be assumed if the ratio of alcohol to glycerine is greater than 10 to 1 by weight.

III. Addition of water and alcohol is recognized by the diminution in the quantity of inorganic matter, especially magnesia, phosphoric acid, and usually potash. Addition of water alone is recognized in the same way.

IV. Scheelization, *i. e.*, addition of glycerine, is assumed if the ratio of glycerine to alcohol exceeds 1 to 6 by weight.

V. The presence of cane sugar is ascertained by a determination of sugar (by Soxhlet's or Knapp's method), before and after inversion.

BEER.

A.—METHODS OF INVESTIGATION.

By beer is to be understood a fermented and still fermenting drink, made from barley (or wheat) malt, hops, and water, and which was fermented by yeast.

I. *Determination of specific gravity.*—For this as well as all other determinations the beer is freed from carbonic acid, as far as possible, by half-filling bottles with it and shaking vigorously. It is then filtered. The specific gravity is then determined either by Westphal's balance or by a picnometer at 15° C.

II. *Determination of extract.*—Seventy-five cubic centimeters of beer are carefully weighed and evaporated in a suitable vessel to 25 cubic centimeters, care being taken to prevent boiling. After cooling water is added until the original weight is reached, and the specific gravity of the liquid taken as in I. The per cent. of extract is obtained from this specific gravity by the use of a table constructed by Dr. Schultz, and is given as "per cent. extract, Schultz."[1]

III. Alcohol is determined by distilling the beer. A picnometer of about 50 cubic centimeters capacity and with a graduated neck is used as a receiver. The picnometer is carefully calibrated. Seventy-five cubic centimeters of beer are distilled until the distillate reaches about the center of the scale on the neck of the picnometer. This is then cooled to 15° C., dried, and weighed, and the alcohol determined by means of Baumber's table.

$$A = \frac{D \cdot d}{g}$$

The percentage of alcohol by weight is to be given. In very acid beers it is necessary to neutralize before distilling.

IV. *Original gravity of wort.*—This may be ascertained, approximately, by doubling the per cent. by weight of alcohol found as above and adding the per cent. of extract. As this procedure is not exact, it may be made more nearly so by using the formula

$$\frac{100\ (E+2.0665\ A)}{100+1.0665\ A}$$

V. Degree of fermentation. This is estimated by using the formula

$$V_1 = 100\left(1 - \frac{E}{c}\right)$$

VI. *Sugar determination.*—This is to be determined directly, in the beer previously freed from carbonic acid, by Soxhlet's method of weighing the reduced copper; 1.13 parts of copper correspond to 1 part anhydrous maltose.

VII. Determination of dextrin is seldom required, and if required is to be performed by Sachsse's method.

VIII. *Nitrogen.*—Twenty to thirty cubic centimeters are evaporated in a Hofmeister "schälchen" or on warm mercury, and the extract burned with soda-lime. The nitrogen may also be determined by Kjeldahl's method.

IX. *Acids.*—(*a*) Total acids: The carbonic acid is driven off from 100 cubic centimeters of beer by heating in beakers for a short time to 40° C., and the beer then titrated with baryta water (one-fifth to one-tenth normal). The saturation point is reached when a drop of the liquid has no longer any action on litmus paper. The acidity is to be given in cubic centimeters normal alkali required for 100 grams beer and as

---

[1] Hilger, p. 123.

grams per cent. of lactic acid. The indication "acidity" or "degree of acidity" is insufficient.

(*b*) Normal beer contains but a very small quantity of acetic acid. The determination of fixed acid in the repeatedly evaporated extract is to be cast aside. The acetic acid produced by souring of the beer is shown by the increase in total acids. A qualitative test of the presence of acetic acid in the distillate from beers containing acetic acid is sufficient. Neutralized beer is to be acidified with phosphoric acid and distilled. Weigert's method is recommended.

X. *Ash.*—Thirty to fifty cubic centimeters of beer are evaporated in a large tarred platinum dish and the extract carefully burned. If the burning takes place slowly, the ash constituents do not fuse together.

XI. *Phosphoric acid.*—This is to be determined in the ash obtained by evaporating and burning in a muffle 50 to 100 cubic centimeters of beer to which not too much baric hydrate has been added. The phosphoric acid is determined in the nitric acid solution of the ash by the molybdenum method.

XII. *Sulphuric acid.*—The direct determination is not permissible. The determination is to be made by using the ash prepared by burning with sodic hydrate and potassic nitrate or baric hydrate and proceeding in the ordinary way.

XIII. *Chlorine.*—This is to be determined in the ash prepared with sodic hydrate.

XIV. *Glycerine.*—Three grams calcic hydrate are added to 50 cubic centimeters of beer, evaporated to a sirupy consistence, about 10 grams coarse sea-sand or marble added and dried. The dry mass is rubbed up, put into a capsule of filter paper, placed in an extraction apparatus, and extracted for six to eight hours with 50 cubic centimeters alcohol. To the light-colored extract at least an equal volume of ether is added, and the solution, after standing awhile, poured into or filtered into a weighed capsule. After evaporating the alcohol and ether, the residue is heated in a drying oven at 100° to 105° C. to a constant loss of weight. In beers rich in extract the ash contained in the glycerine can be weighed and subtracted. In case the glycerine contains sugar this can be determined by Soxhlet's method and subtracted.

XV. Hop substitutes are to be determined by Dragendorff's method. Picric acid is to be determined by Fleck's method. In examining for alkaloids, check experiments with pure beer must in all cases be made.

XVI. *Sulphites.*—One hundred cubic centimeters of beer are distilled after the addition of phosphoric acid and the distillate conducted into iodine solution. After one-third has distilled over, the iodine-colored distillate is acidified with hydrochloric acid and baric chloride added. If sulphites were not contained in the beer no precipitate is observed, but at the utmost a turbidity.

XVII. *Salicylic acid.*—This may be shown qualitatively by shaking with ether, chloroform, or benzine. The solution is allowed to evaporate, the residue dissolved in water, and a very dilute solution of ferric chloride added. The addition of too much acid and too violent shaking is to be avoided. The smallest trace of salicylic acid may also be shown by dialysis, as it passes very readily through membrane.

NOTE.—All the results of an investigation are to be stated in percentages by weight.

B.—METHODS OF JUDGING PURITY OF BEERS.

I. It is unjust to demand in a fermented beer an exact ratio of alcohol to extract, as the brewer can not regulate the degree of fermentation within narrow limits. As a rule, Bavarian draught and lager beers contain from 1.5 to 2 parts of extract for each part of alcohol, but a smaller proportion of extract would not necessarily prove the addition of alcohol or glucose (the former to the beer, the latter to the wort).

II. The degree of fermentation of a beer must be such that at least 48 per cent. of the original extract has undergone fermentation.

III. If glucose or other bodies poor in nitrogen have been used in appreciable quantity as substitutes for malt, the nitrogen contents of the beer extract will fall below 0.65 per cent.

IV. The acidity of a beer should not be greater than 3 cubic centimeters normal alkali to 100 cubic centimeters beer. Acidity of less than 1.2 cubic centimeters normal alkali to 100 cubic centimeters beer indicates previous neutralization. If the acids are composed principally of lactic acid a larger quantity may be present.

V. The ash of normal beer is not above 0.3 gram to 100 grams beer.

VI. The amounts of phosphoric and sulphuric acids and chlorine in beer extract vary within such wide limits, that their determination signifies nothing as to the purity of the beer.

VII. The amount of glycerine in pure beer is not greater than 0.25 gram to 100 grams beer.

VIII. The following methods of clarifying beer are legal: (a) Filtration. (b) Well-boiled hazel or beech shavings. (c) Isinglass.

IX. The following methods of preserving beer are legal: (a) Carbonic acid. (b) Pasteurizing. (c) Salicylic acid; this only for beers intended for export to countries where the use of salicylic acid is not forbidden by law.

NOTE.—The preceding methods are also to be used in the examination of imported beer.

C.—ADMINISTRATIVE NOTE.

It is absolutely necessary that the beer be preserved in well-corked green-glass bottles. Stone jugs and such vessels are not to be used.

The beer samples are to be protected from light and kept at a low temperature.

Care in making tests is, above all, necessary.

OFFICERS AND REPORTERS OF THE ASSOCIATION OF OFFICIAL AGRICULTURAL CHEMISTS OF THE UNITED STATES FOR 1887-1888.

PRESIDENT,

Prof. JOHN A. MYERS, Director of the West Virginia Agricultural Experiment Station

VICE-PRESIDENT,

Prof. M. A. SCOVELL, Director of the Kentucky Agricultural Experiment Station.

SECRETARY,

CLIFFORD RICHARDSON, Chemist to the District of Columbia and Honorary Assistant Chemist United States Department of Agriculture.

EXECUTIVE COMMITTEE.

Dr. H. W. WILEY, of Washington.
Dr. WILLIAM FREAR, of Pennsylvania.

REPORTERS.

Phosphoric acid.—Dr. Richard H. Gaines, of Richmond, Va.

Potash.—Dr. E. H. Jenkins, of New Haven, Conn.

Nitrogen.—Prof. M. A. Scovell, of Lexington, Ky.

Cattle foods.—Prof. G. C. Caldwell, of Ithica, N. Y.

Dairy products.—Dr. H. W. Wiley, of Washington, D. C.

Fermented liquors.—Prof. W. B. Rising, of Berkeley, Cal.

Sugar.—Prof. W. C. Stubbs, of Kenner, La.

CONSTITUTION OF THE ASSOCIATION OF OFFICIAL AGRICULTURAL CHEMISTS.

(1) This association shall be known as the Association of Official Agricultural Chemists in the United States. The objects shall be (1) to secure uniformity and accuracy in the methods, results, and modes of statements of analysis of fertilizers, soils, cattle foods, dairy products, and other materials connected with agricultural industry; (2) to afford opportunity for the discussion of matters of interest to agricultural chemists.

(2) Analytical chemists connected with the United States Department of Agriculture, or with any State or national agricultural experiment station or agricultural college, or with any State or national institution or body charged with official control of the materials named in section 1, shall alone be eligible to membership, and one such representative for each of these institutions or boards, when properly accredited, shall be entitled to a vote in the association. Only such chemists as are connected with institutions exercising official fertilizer control shall vote on questions involving methods of analyzing fertilizers. Any person eligible to membership may become a member at any meeting of the association by presenting proper credentials and signing this constitution. All members of the association who lose their right to such membership by retiring from positions indicated as requisite for membership shall be entitled to become honorary members and to all privileges of membership save the right to hold office and vote. All analytical chemists and others interested in the objects of the association may attend its meetings and take part in its discussions, but shall have no vote in the association.

(3) The officers of the association shall consist of a president, vice-president, and secretary, who shall also officiate as treasurer; and these officers, together with two other members to be elected by the association, shall constitute the executive committee. When any officer ceases to be a member by reason of withdrawing from a department or board whose members are eligible to membership, his office shall be considered vacant, and a successor may be appointed by the executive committee to continue in office till the annual meeting next following.

(4) There shall be appointed by the president, at the regular annual meeting, a reporter for each of the subjects to be considered by the association.

It shall be the duty of these reporters to prepare and distribute samples and standard reagents to members of the association and others desiring the same; to furnish, blanks for tabulating analyses, and to present at the annual meeting the results of work done, discussion thereof, and recommendations of methods to be followed.

(5) The special duties of the officers of the association shall be further defined, when necessary, by the executive committee.

(6) The annual meeting of this association shall be held at such place as shall be decided by the association, and at such time as shall be decided by the executive committee, and announced at least three months before the time of meeting.

(7) Special meetings shall be called by the executive committee when in its judgment it shall be necessary, or on the written request of five members; and at any meeting, regular or special, seven enrolled members entitled to vote shall constitute a quorum for the transaction of business.

(8) The executive committee shall confer with the official boards represented with reference to the payment of expenses connected with the meetings and publication of the proceedings of the association.

(9) All proposed alterations or amendments to this constitution shall be referred to a select committee of three at a regular meeting, and after report from such committee may be adopted by a vote of two-thirds of the members present and entitled to vote.

96